Book A1

Math Ties

Problem Solving, Logic Teasers, and Math Puzzles

All "Tied" to the Math Curriculum

SERIES TITLES

Math Ties Book A1 • Math Ties Book B1

written by

Terri Husted

illustrated by

Rob Gallerani

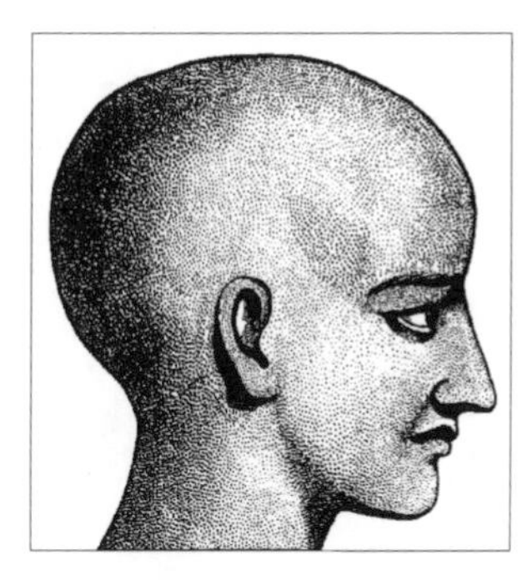

CRITICAL THINKING BOOKS & SOFTWARE
www.CriticalThinking.com
P.O. Box 448 • Pacific Grove • CA 93950-0448
Phone 800-458-4849 • FAX 831-393-3277
ISBN 0-89455-670-3
Printed in the United States of America

TABLE OF CONTENTS

To Dave and my children,
Alexis and Venissa

Introduction

Math Ties Book A1 is a classroom-tested approach to help you teach problem solving to *all* your students (grades 4-6), regardless of ability. With a track record of 16 successful years, the method provides an organized step-by-step approach to incorporating problem solving on a regular basis in the classroom, and it includes a collection of problems your students will love to solve.

As a beginning teacher, I realized that learning problem solving is far too complex a process to be developed by giving a short unit or occasional challenge problems. Many students, especially the math anxious or those with previous bad experiences in math, would improve their math skills but make little progress in problem solving. Many had trouble retaining strategies from one year to the next, or even from one lesson to the next. Of the many books available on problem solving, however, none give a methodology for incorporating problem solving as a continuous and integral unit in the classroom. Also, finding a good collection of problems organized by curriculum themes is nearly impossible without spending an enormous amount of money.

To resolve these issues, I have developed the *Math Ties* problem-solving method. One day a week throughout the entire year is devoted to teaching problem-solving strategies by letting students work on critical-thinking problems in mathematics. I have incorporated the method with classes as large as 36 (for classes larger than 30, it helps to have an assistant). During the rest of the week, I use a variety of methods including lectures and other activities that promote critical thinking, but on problem-solving day we use *Math Ties* only. My students call it "problem-solving day,"— and they love it.

What is problem solving?

To describe this teaching approach, I must first define what problem solving is in mathematics. Problem solving must involve a series of critical-thinking skills. A good problem in mathematics requires some thought process beyond a quick arithmetic solution. It also involves more than just working with word problems, although there are many word problems that are excellent critical-thinking problems. A "problem-solving" problem can be a puzzle, a game, a brainteaser, a logic problem, or any problem that engages the reader in using a strategy. A good problem enhances the concept you have been teaching, promotes discussion, and allows students to enjoy and gain a deeper meaning of mathematics.

The key to teaching critical-thinking skills is that the problem solving must happen as often as possible and must be part of an organized, well-thought-out approach. Learning how to problem solve is a very individual and personal process. Teaching problem solving requires step-by-step modeling, gradual presentation of strategies, many opportunities for students to practice the same strategy with many different types of problems, constant practice in a nonthreatening environment, follow-up time for reflection on each problem, and—that rare occurrence in most classrooms—time allowed for long-term growth (not all students learn at the same rate). Moreover, problem solving works best when it is taught in the context of the topic being learned and should include, whenever possible, a multicultural component.

The goals of *Math Ties* are for all students to (1) become more proficient in using problem-solving strategies (see pp. xi, xii);

(2) attempt all problems without giving up right away; (3) gain confidence in their math ability and become more independent; (4) see connections between problems, between topics, and between disciplines; and (5) see solving problems as a challenge and not as an overwhelming obstacle, whether in math class or in everyday life.

One problem-solving strategy is the use of manipulatives. My room is equipped with blocks, tiles, construction paper, markers, scissors, glue, toothpicks, scrap paper, rulers, compasses, calculators, and large poster paper so that students can diagram the problem, make cutouts, and move "people" about (in bridge problems, etc.). I don't have expensive manipulatives; many of my students like creating their own while solving a problem. I often encourage students to dramatize the problem because acting out a problem is fun and helps students see parts of the problem not previously seen.

Whether you use *Math Ties* once a week or more, what is most important is that you do it consistently throughout the year. This may seem impossible with so much curriculum to cover, but most of the problems you will use are enrichment problems from your curriculum that will enhance and expand what you are already teaching. The *Math Ties* problem-solving day will be the highlight of the week for you and your students. With reinforcement, modeling of strategies, and honest praise, students learn to explore their critical-thinking capabilities as they never have before, and they learn to love mathematics.

At the beginning of the year, I give my students a Math Myths survey (see p. 77) to understand their preconceived notions about math. I constantly remind my students that mathematics does not belong to the well-known scientists or the brightest minds; mathematics belongs to every one of us. Becoming good at problem solving takes time and practice, but it requires more than a good collection of problems. It also takes clear expectations, lots of planning, and a well-structured environment that is warm and caring and where students can be free from fear of failure.

Choosing the problems

I choose my problems according to curriculum themes (see Table of Contents, p. iii, Matrix of Problem Solving Concepts, p. 78, and Matrix of Problem Solving Strategies, p. 83). I have numbered the problems in the matrix only to help you determine what concepts and problem-solving strategies are covered by each problem. You can use the problems in any order that best fits your curriculum and the needs of your students. However, during the first week I usually begin with logic problems and classic brainteasers so I can learn more about the students' math abilities, their group work behavior, their level of math anxiety, and their social skills. Kin problems are also great problems to start with because students love them and they clearly demonstrate the importance of making diagrams.

This collection includes the most successful problems I have used, adapted for upper elementary students. Variations of these problems can be found in brainteaser books, history of math books, enrichment sections of textbooks, math contests, etc. Many of the problems are contributions made by fellow teachers, students, parents, and even friends and neighbors. Some problems work better with certain levels, and it is important to document those results. Many younger students and students who are not very motivated in math do better if they start with problems that require manipulatives, dramatization, cutting, pasting, and coloring. It is the task of a good math teacher to get all students to engage in mathematical discussions and to encourage students to use mathematical vocabulary whenever possible. Some students, at first, are not comfortable with problems that require sophisticated explanations and open discussion among peers. Be patient. Most students need to gain some confidence before engaging in any activity that requires risk-taking. Some of the best thinking strategies I have seen have come from those labeled "lower-level" students. It is very important to validate each student's response and effort.

How to get started

Set up your classroom in groups of 2 to 4 students (3–4 is best). Any student who wants to work alone should be gently encouraged to work in a group. Groups larger than 4 are often too large. I let students choose their own group, but I do make changes as needed.

Math Ties includes enough material for at least two problems per session for a 36-week school year. You should augment this number to four or five problems per session by collecting additional problems throughout the year. Select the problems you will use for a one-period class and cut them separately in strips, allowing one problem per student.

The beginning of class is a good time to discuss a problem-solving strategy. I have included a problem-solving concept map that you can use with your students to model how to solve a problem. Make copies of the problem-solving strategies and the concept map, and have students refer to them during each session. I have a large section of my wall devoted to problem solving where I can post the strategies, the group rules, and many samples of the students' work.

The class comes in on problem-solving day and waits to hear me read the first (or my favorite) problem aloud. During this quiet time, I review group rules and alert them to try certain strategies that may be most useful for that day's session. Each group is then asked to choose a team leader. Reading the first problem aloud promotes interest and gets the students excited about starting. I hand out a copy of the first problem to each person in the room. The rest of the problems are spread out on a table (one copy of each problem for every student). This consistent method of starting helps maintain a structured environment.

The job of the team leader is to keep the group on task and to go to the table to get a copy of the next problem for each person in the group. Allow the team leader to get only one problem at a time; otherwise, the members of the group are likely to work on several different problems at the same time, and that's the end of cooperation and good discussion among team members. Stress that it is not necessary to do *all* the problems—it is more important to work carefully on each one. Once they have experienced problems with more than one answer (see The Chicken, the Corn, and the Fox, p. 1) and found that the first answer is not usually the correct one, they will become more cautious. Soon they will start reading each problem more carefully, which is one of the goals of problem solving.

To grade or not to grade

Becoming good at problem solving takes time, but it is a skill that all students can develop with practice. Some students have a negative attitude towards math problems, and many (including adults) strongly dislike word problems. Many students have had bad experiences with word problems because of poor reading skills, poor math skills, attitudes shared by peers (or even parents), lack of practice, and previous failures. Many students fear getting stuck and appearing "stupid" in front of their peers. Some avoid even starting a problem because they are convinced that others are always better than they are at solving problems. They fear the time pressure and a low or failing grade. Nevertheless, many teachers are reluctant to try anything without attaching a grade to the activity; they fear that students won't work unless they "have to." I strongly recommend that problem solving *not* be graded. Students will take risks, enjoy the activities, and seek you out for more problems if the fear of a grade is lifted. As long as the problem-solving experience is free from the pressures of a grade, students will develop many or all of the problem-solving strategies in my list, given time and experience. I am convinced that a nonthreatening environment in the classroom is crucial to the success of developing good critical-thinking skills in mathematics.

You can evaluate each student's progress by keeping an index card on each student, recording notes on progress, or giving extra credit for problems done individually at home. You can give extra credit to those students who expand, rewrite, or diagram the problem on a poster, etc. It may be difficult to walk around taking notes on students' responses and progress. Therefore, you may occasionally want to use the Problem Solving Interpretation (p. xiii) and Analysis of Solutions (p. xiv) forms (with a problem of your choice or one chosen by the student) in order to analyze how students attack problems. You can hand out these sheets the last fifteen minutes of class and ask students to pick one problem to analyze. You will learn an enormous amount about mathematics acquisition and critical thinking by listening to students' comments and explanations.

Portfolios offer another evaluation method. Have students create a portfolio of problems they choose according to several categories (see Appendix A for the student forms). The worksheets are designed to encourage students to reflect on their growth. Keep their problems in a folder so they can take them home, and let them choose the problems for their final portfolio to turn in to you at a later time. Be careful not to let the portfolio work or any critical-thinking evaluation sheets become the goal of the problem-solving unit. The goal is for your students to experience many types of problems, enjoy problem solving, and become stronger at it.

The teacher's role during problem solving

The role of the teacher in a problem-solving session is that of facilitator, role model, and guide. You must make sure that the problem-solving rules are being followed. The level of noise in the classroom is bound to be higher, but students will work and follow the rules if you set clear expectations and review the rules periodically. One way to avoid students' calling out is to use the "red and green" cups. Tape together the bottoms of two cups, one green and one red, so that students can put the red side up when they need help and keep the green side up when they do not need help. One "red and green" cup is used per group. You can experiment creating some other device that students can use to let you know they need help without having to call you or raise their hands. Try not to answer any questions for the first ten minutes of your problem-solving session; your silence encourages the group to check with each other. Also, if you are unable to come over right away, students will keep on working and often end up not needing help. Try not to answer questions unless everyone in the group has been consulted and the entire group needs a hint in order to continue working.

When the group comes up with an answer, expect an explanation. You need not always tell them if they are correct. Eventually, they won't check for approval once they are satisfied that their answer is correct. Some problems have more than one answer, so encourage your students to find other solutions. For some problems, the group's final answer may not be the most efficient. Always accept and praise original solutions, but ask students to stick to the restrictions given by the problem.

Classroom discipline

There are several ways to maintain good classroom discipline when doing a problem-solving session. The classroom will not be quiet, but it should not be so noisy that students have problems communicating within their own group. Students must learn to stay in their group and not visit other groups. Here are some ways to avoid discipline problems:

1. Frequently review rules and expectations at the beginning of the session.
2. Allow only the team leader to get up. Don't allow visiting between groups.
3. Stop all students a few minutes before class is over so they can clean up, put problems in a folder (or you may collect them from each group yourself), and straighten up the room.
4. Allow students to work with some of their friends. Students forced to sit with someone they strongly dislike can end up hating group work. Social conflicts are more frequent in middle school, so use your best judgment and, above all, be fair and consistent. Some students need to work together with the same set of students for a few sessions to feel more comfortable. Change groups only when there is good reason to. Teachers may disagree on how to set up groups, so try several approaches and choose what works best for you.

Don't expect perfection the first few times you try this approach; it will take time. If you are enthusiastic about problem-solving day with *Math Ties*, your students will be as well.

What to do with the problems that are solved

Students can keep all their solved problems in a neat pile on their group table with their names on them and turn the problems in at the end of the period. Hang as many problems as possible on the wall. Encourage students to extend the problem, diagram results, or create a poster of the problem for extra credit. If a group has finished all the problems, they can start on their homework or you may give them an extra problem to solve. Many students enjoy taking problems home for their parents to solve. Don't forget to give feedback to the parents and also hang the results in your classroom. Follow-up discussions are very important; a good time for the discussion is the next day at the beginning of class.

Above all, praise and praise

With continuous reinforcement throughout the year, a warm and caring environment, freedom from fear of failure, great problems to solve, and constant honest praise, students will grow at problem solving and learn to love mathematics.

GROUP RULES FOR PROBLEM SOLVING

1. Stay in groups of 2, 3, or 4.
2. Everyone must participate.
3. Work together as a group. Do not let anyone in your group get behind or ahead.
4. Use the problem-solving strategies posted.
5. Help those in your group who do not understand, but do not do the work for them.
6. Be a good listener. Do not make fun of anyone's ideas.
7. Do not ask for help until everyone in your group has been consulted and has no idea how to proceed. Keep working until I get to you.
8. Remember, there is often more than one way to solve a problem.
9. Team leaders: Keep your group on task and come get the next problem when everyone is ready.
10. It's OK not to do all the problems. Relax and enjoy math!

PROBLEM-SOLVING STRATEGIES

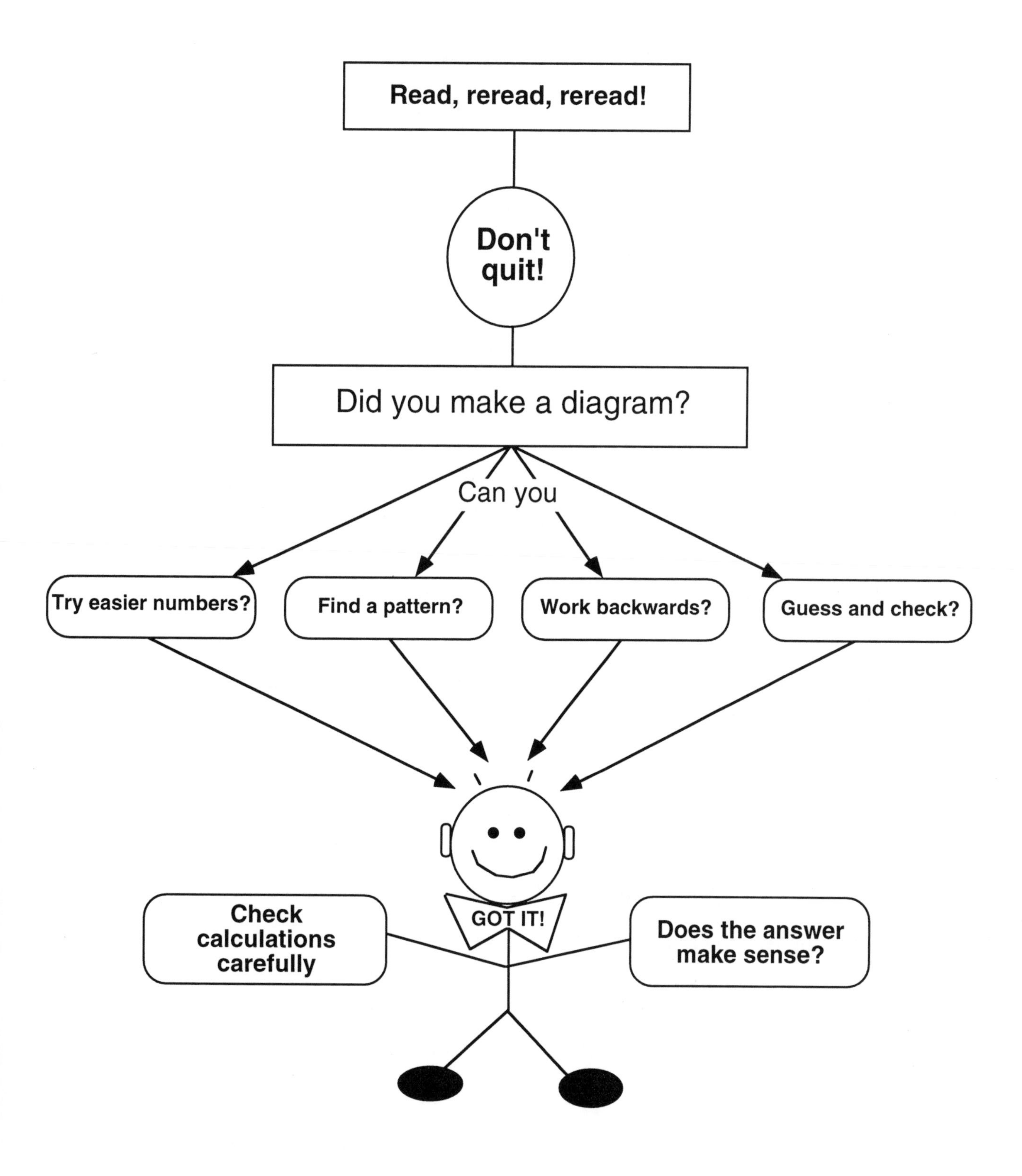

PROBLEM-SOLVING STRATEGIES

Read, reread, and reread

1. Say the problem aloud. Explain it in your own words to someone else.
2. Before you start, be sure you understand what is being asked.
3. Define any words you don't know.
4. Look for key words, but be sure to read in context.
5. Be careful with unnecessary information.
6. Think of a similar problem you've solved before.
7. Don't quit! It's okay to put the problem aside and come back to it later.

Attack the problem

8. Use manipulatives or dramatize the problem.
9. Make a drawing, chart, graph, or model.
10. Take a guess and check your answer.
11. Find a pattern or formula.
12. Try easier numbers. Sometimes this helps you find a pattern.
13. Work backwards.
14. If the work seems overwhelming, there is usually a much easier way to solve it.

Check your answer

15. Check your calculations carefully. Did you use the correct unit?
16. Did you answer the question being asked?
17. Read the problem again; does the answer make sense?
18. Are there any other possible answers you did not consider?

PROBLEM-SOLVING INTERPRETATION

Problem Title ______________________________

State the problem in your own words:

Diagram the problem:

ANALYSIS OF SOLUTIONS

Problem Title ______________________________

I tried these solutions, and they didn't work:	Why not:

THIS SOLUTION WORKED!

LOGIC PROBLEMS

To the teacher

Most of the problems in this section are best solved by manipulatives or by dramatization.

There are many variations of The Chicken, the Corn, and the Fox. Western versions also include a wolf, a goat, and a cabbage. African versions include a jackal, a goat, and a bundle of hay or a cheetah, a fowl, and some rice. While the logic puzzles are similar, they vary as to how many items are carried across the river at a time. Share with students the fact that many cultures enjoy "crossing-the-river" problems. You can find more of them in *Math Ties* Book B1.

The Blocks Puzzle is a good problem to do after reviewing the terms *parallel lines*, *pentagon*, and *trapezoid*. (If you haven't taught these terms, make sure that you identify their names and show the shapes on the board before handing out the problem.) I have provided a grid that students may use in eliminating figures that are not opposite. Using the grid is a good introduction to the use of matrices, which will be needed in more difficult logic problems.

1 The Chicken, the Corn, and the Fox

A man comes to a river he must cross. He needs to take his chicken, his corn, and his fox across, but the boat can hold only one item besides himself. He cannot leave the fox alone with the chicken or the fox will eat the chicken, and he cannot leave the chicken alone with the corn or the chicken will eat the corn. How can he take them across? (The fox does not like corn.)

2 A Family Problem

A family wants to cross a river in a boat that holds no more than 250 lbs. The husband weighs 160 lbs. His wife weighs 130 lbs. The daughter weighs 85 lbs., and the son weighs 115 lbs. How can they all get across in the fewest number of trips?

How will the answer change if you cannot leave the smallest child, the daughter, alone at any time?

3 Truthful?

A student tells his teacher, "I always lie. I never tell the truth." What is wrong with this statement?

The Liars Club

The members of the Liars Club never tell the truth. After the school's spelling bee, one of them said, "Katie finished first. Alex beat Katie. Susie beat Tom, and Alex came in last." In what order did the students really finish the spelling bee?

5 Liza at the Pond

Liza has gone to the pond with a large bucket and two jars, none of which have any markings. One jar holds exactly 5 cups, and the other one holds exactly 3 cups. Liza needs to bring home exactly 1 quart of water (1 quart = 4 cups). How can she do it?

6 Line Up!

Juan is younger than Tammy, and he is older and shorter than Peter. Tammy is taller and younger than Carmen, yet Carmen is taller than Peter. Cut out the name cards, and then follow the directions below them.

Juan	Carmen	Peter	Tammy

Juan	Carmen	Peter	Tammy

Put the friends in order from youngest to oldest.

Put the friends in order from shortest to tallest.

7 The Blocks Puzzle

Each drawing represents a different view of the *same* block:

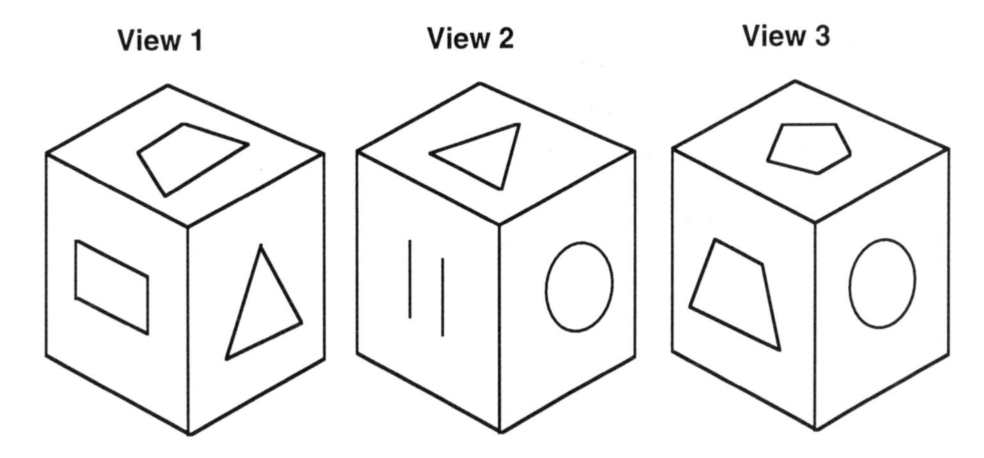

What's on the opposite side of each figure? (Finish the grid below by crossing out the shapes that you know are not opposite; put a O in the box for shapes that are opposite.)

	TRAPEZOID	RECTANGLE	TRIANGLE	PARALLEL LINES	CIRCLE	PENTAGON
TRAPEZOID	X					
RECTANGLE		X				
TRIANGLE			X			
PARALLEL LINES				X		
CIRCLE					X	
PENTAGON						X

Figure:	**Opposite figure:**
trapezoid	________________
rectangle	________________
triangle	________________
parallel lines	________________
circle	________________
pentagon	________________

WHOLE NUMBERS

To the teacher

Whole-number problems provide students with practice in whole numbers as they work with problem-solving strategies.

Encourage your students to use a chart or diagram for most of these problems. In The Missing Symbols, encourage them to think of the prime factors of 100.

Whenever a problem calls for it, have students give an explanation. It's a good way to get them to use writing while doing math.

8 The Missing Symbols

Replace each box with one of the following four symbols so that you make 100:

$+ \quad \bullet \quad (\quad)$

$$\square\ 1\ \square\ 2\ \square\ 3\ \square\ 4\ \square\ 5\ = 100$$

9 The Missing Steps

Julia and Maria live in an apartment building. From the first floor to the third floor there are 52 steps. Julia lives on the first floor. How many steps would she climb to get to Maria's apartment, which is on the sixth floor? Assume the same number of steps between floors. Draw a picture to help you.

10 The Cool Company

In the hot summer, the Cool Company sells lots of ice cubes. They sell them only in bags of 5 pounds, 12 pounds, 20 pounds, and 35 pounds. A customer wants to buy exactly 129 pounds. How can the Cool Company meet the customer's request?

11 The Slippery Well

A frog lives in a slippery well 20 feet deep. Every day it hops up 5 feet but then slips down 4 feet. In how many days does it get out of the well? (The frog starts at the bottom of the well.)*

*Once the frog is outside of the well, it will happily hop away!

12 The School Math Competition

At a school math competition, students were asked to solve 10 problems. Each question answered correctly received 5 points, and each question answered incorrectly resulted in a loss of 3 points. How many problems were correctly solved by a student who received a final score of 34 points?

13 The Toy Factory

A toy maker takes 42 minutes to make a new doll and 1 1/2 hours to make a new bike. Assuming only one toy can be made at a time, how long would it take the toy maker to complete an order for five bikes and 10 dolls?

If he started at 6:30 A.M. and worked without stopping, at what time would the toy maker complete this order?

14 Todd's Turtle

Todd bought a turtle for $15. His mom was not too happy about the turtle, so Todd sold it for $20. Later, his mom changed her mind and Todd bought it back for $23. A week later, Todd sold it to his friend Ruben for $30. Did Todd make or lose any money? If so, how much?

15 The Hockey Camp

A hockey camp advertises a special deal for the winter break, as shown below.

Plan A: Family Groups—$400 a week.

Plan B: Individuals—First person $250 a week
Each additional person: $60 a week.

Mike and his three brothers, Dave, Alex and Brian, all want to go. Which plan is better for them? Explain your answer.

If Brian decides he can't go, which plan is better for Alex, Dave, and Mike? Explain your answer.

NUMBER THEORY

To the teacher

This section includes problems that encourage students to think of patterns and gain an appreciation for working with numbers. In The Salary Problem, The Never-ending Joke, and Going to the Beach, students often think that their last exponent result (i.e., 2^5) is the final answer; they need to add the amounts found at each stage to get the total. The Narrow Staircase is an example that illustrates the importance of adding the subtotal after each step.

Going to the Beach is based on the problem of St. Ives, which dates back to Egyptian times. It would be fun for your students to research other old problems that have been handed down over the centuries.

The Famous Odd Numbers is best done together with The Proud, Perfect Squares so that students can appreciate the pattern that is created by adding the odd numbers. You may also find it helpful to connect The Proud, Perfect Squares with other work on areas of perfect squares.

16 The Never-ending Joke

Tammy tells a joke to two friends.

The next minute, each of her two friends tells the same joke to two other friends; the next minute, each of those friends tells the same joke to two other friends; this keeps going on until 5 minutes have passed and the teacher begins to pull her hair and yells "Stop!" How many friends know the joke after 5 minutes? (Don't forget to add Tammy.)

Now do you see why rumors spread so quickly?

The Presidents

Five presidents attend an important meeting. They all shake hands with one another. How many handshakes are there? (Use pictures or a chart to help you.)

If there were 10 presidents at this meeting, how many handshakes would there be?

18 Mail in Remote Land

On a remote island, mail arrives every 3 days, groceries arrive every 4 days, and a shipment of clothes arrives every 5 days. How often do groceries, mail, and clothes arrive on the same day?

19 TV Cartoons

Two TV cartoons start on different channels every day at 7A.M. Froggy, the Mathematician runs for 90 minutes with no commercials then starts again. The Hunt for Cinderella's Ruler runs for 55 minutes with 5 minutes of commercials, then starts again. Assuming that both shows run again and again in this manner all day, what is the next time of day when they start at the same time?

The Salary Problem

You've just been offered a job, and you have a choice of salary. Would you rather have $1200 a month or 1 cent the first day, 2 cents the second day, 4 cents the third day, 8 cents the fourth day, etc., doubling each day, for 30 days?

21 The Narrow Staircase

A very narrow staircase is made from blocks (cubes) and goes up 5 steps as shown below. How many more bricks are needed to make the staircase 10 steps?

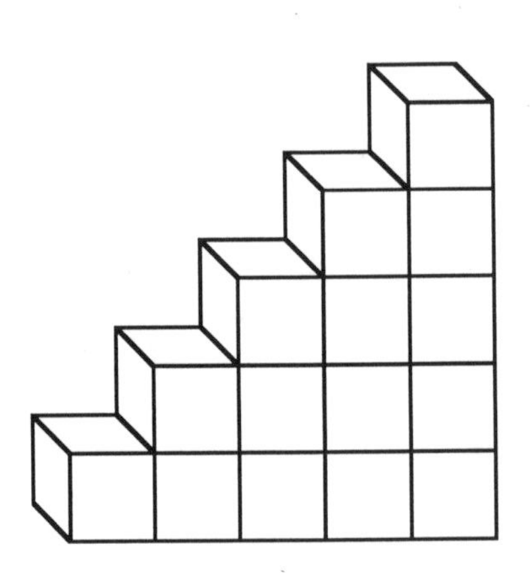

22 The Proud, Perfect Squares

The numbers 1, 4, 9, and 16 are called perfect squares. Here's why:

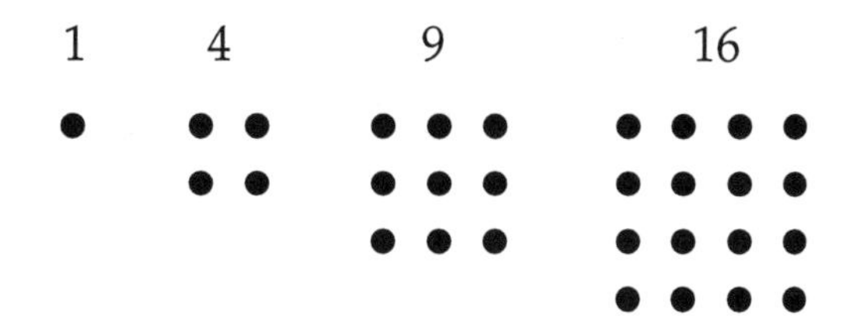

Can you find the next number that is a perfect square and make a diagram for it?

Can you predict the next two numbers (after the last one you found) that are perfect squares?

23 The Famous Odd Numbers

Here are the famous odd numbers: 1, 3, 5, 7, 9, 11, 13, 15, 17, 19...

Find the sums:

1 + 3 ______

1 + 3 + 5 ______

1 + 3 + 5 + 7 ______

1+ 3 + 5 + 7 + 9 ______

1+ 3 + 5 + 7 + 9 + 11 ______

Continue the pattern. Can you figure out what is happening?

24 The Bag of Marbles

You have a bag of marbles. If you count them by 2s, you get a remainder of 1. If you count them by 3s, you get a remainder of 2. If you count them by 4s, you get a remainder of 3. However, if you count them by 5s, you get no remainder. How many marbles do you have? (Use the hundreds chart to help you.)

1	2	3	4	5	6	7	8	9	10
11	12	13	14	15	16	17	18	19	20
21	22	23	24	25	26	27	28	29	30
31	32	33	34	35	36	37	38	39	40
41	42	43	44	45	46	47	48	49	50
51	52	53	54	55	56	57	58	59	60
61	62	63	64	65	66	67	68	69	70
71	72	73	74	75	76	77	78	79	80
81	82	83	84	85	86	87	88	89	90
91	92	93	94	95	96	97	98	99	100

25 Going to the Beach

As Sarah was going to the beach, she saw a woman with two children. Each child carried two bags. Each bag had two buckets. Each bucket had two shovels. Each shovel had two shells. If they were all going to the beach, how many children, bags, buckets, shovels, and shells were going to the beach?

FRACTIONS

To the teacher

Problems in this section will help students diagram their understanding of fractions and fraction operations.

Some students think they are not allowed to cancel the numerator and the denominator of the same fraction (some don't recognize it as reducing because they usually see cross cancelling along diagonal lines when multiplying fractions). The Mystery Fraction helps to correct this problem by giving students practice in cancelling the numerator and denominator of the same fraction.

Fraction Diagram

Make a diagram to show that 2/3 of 9 is 6.

Diagram of Difference

Make two diagrams to show that you know the difference between 10 ÷ 2 and 10 ÷ 1/2. (It's not the same answer!)

28 Wanted!

Two fractions are hiding somewhere on the number line. They are both larger than 1/8 but smaller than 1/4, yet they are different fractions and they are wearing different denominators. Can you find two fractions that fit this description?

29 The Mystery Fraction

Can you find the mystery fraction that makes this product come out to 1? (A product is the result of multiplication.)

$$\frac{9}{18} \times \frac{2}{3} \times \frac{4}{8} \times \frac{7}{14} \times \boxed{\frac{\quad}{\quad}} = 1$$

30 Ms. Kaplan's Musicals

Ms. Kaplan, the drama director, is putting on another musical for her school. It consists of the segments that follow. Act I: 20 minutes, Act II: 33 1/2 minutes, Act III: 32 1/2 minutes, Act IV: 20 minutes—and a 15-minute intermission between Acts II and III. To change the scenery, there must also be 3 1/2 minutes between Acts I and II and between Acts III and IV.

How many hours and minutes are there from the beginning of the musical to the end?

If the musical must end at 10:30 P.M. exactly, what is the latest time it could start?

31 The Bouncing Ball

A ball is dropped from a building 100 feet high. Each time it bounces, it bounces halfway up. Finish the drawing by filling in the bounce heights.

How far up does it rise right after the 5th bounce?

In all, how far has the ball traveled when it hits the ground at bounce 6?

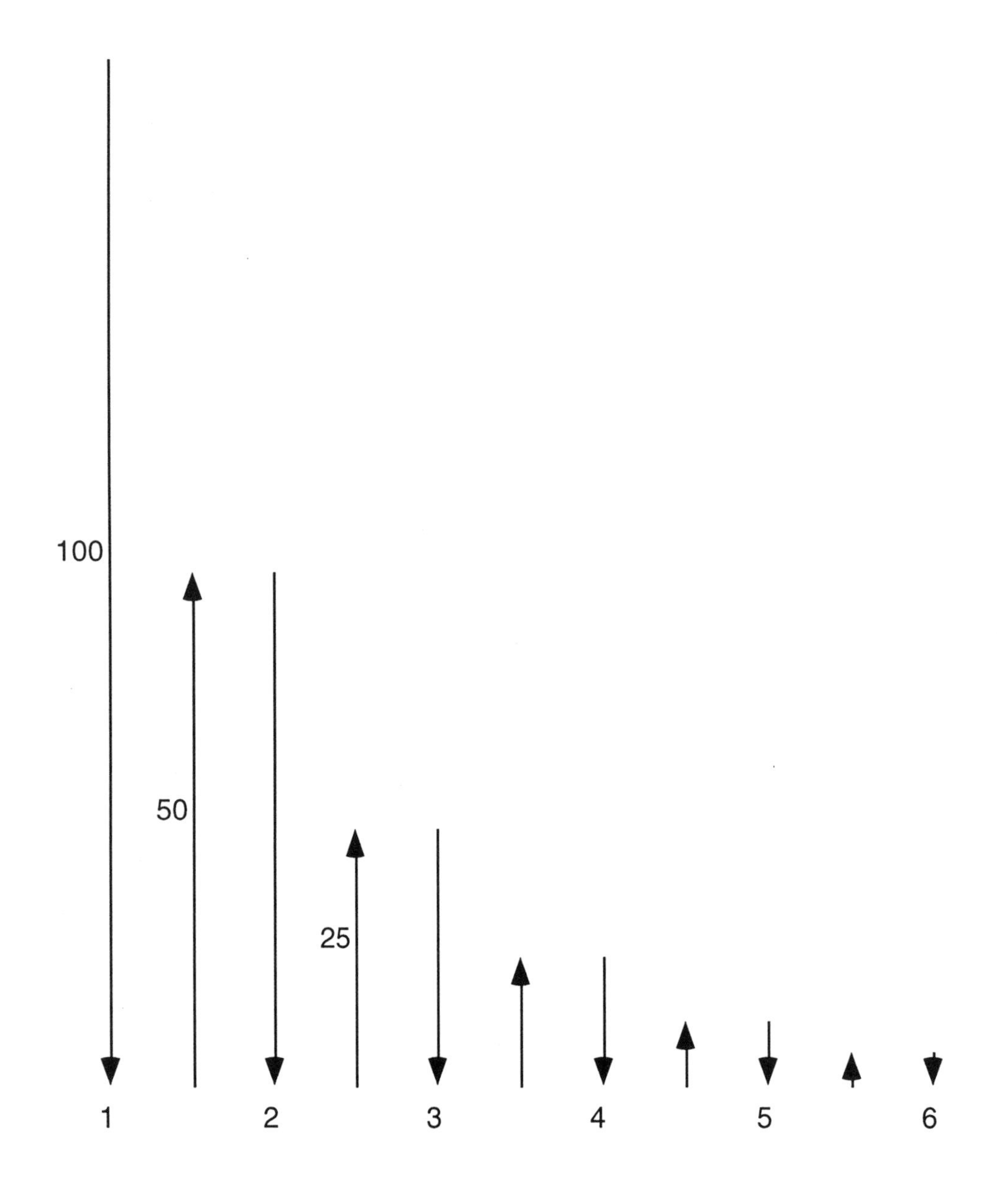

32 The Tricky Fence

A fence has 10 posts equally spaced. From the outside edge of the first post to the outside edge of the last post there are 8 feet. Each post is 3 inches wide. How wide is each space? (Make a picture!)

DECIMALS

To the teacher

Many of the problems in this section were chosen to help you correct some of the most common mistakes made by students who are working with decimals. In Your Turn to Explain, it is a common mistake for students to bring up the remainder as the decimal. In the problem Who's Taller? students often do not see any difference between 5.6 feet and 5 feet 6 inches. One of the reasons for these misconceptions is the fact that we don't often teach fractions together with decimals. Once students are proficient with fractions and have also worked with decimals, we must go back and work with both concepts together so that students start seeing the connections between fractions and decimals.

Place Value Game

Cut out the squares with digits 0, 1, 2, 3, 4, 5, 6, 7, 8, and 9, and place them in a bag. Take turns drawing numbers with one other person; each time you draw, make sure you put the number back in the bag. See who can make the biggest number by writing each digit in one box.

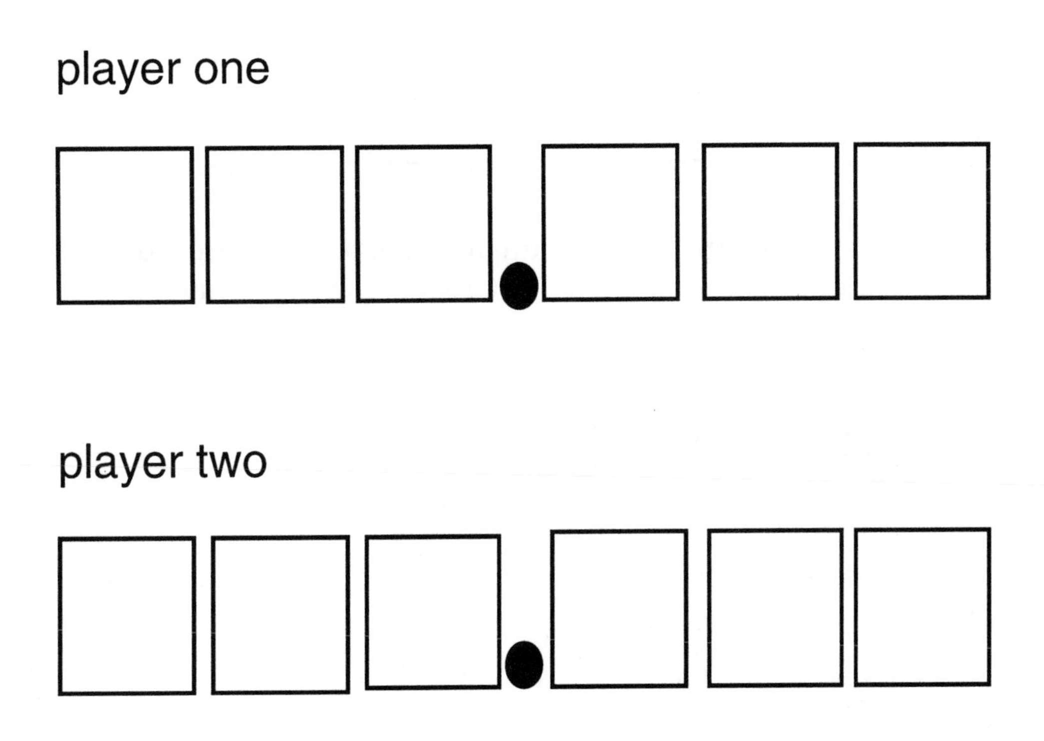

Cut out the squares below:

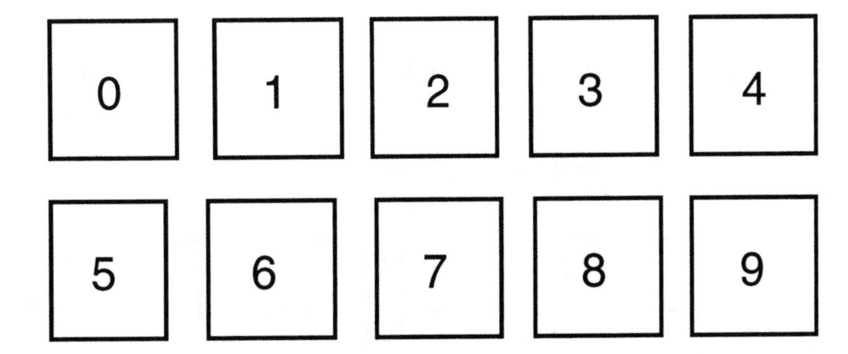

34 Which is Bigger?

Jim saw this sign:

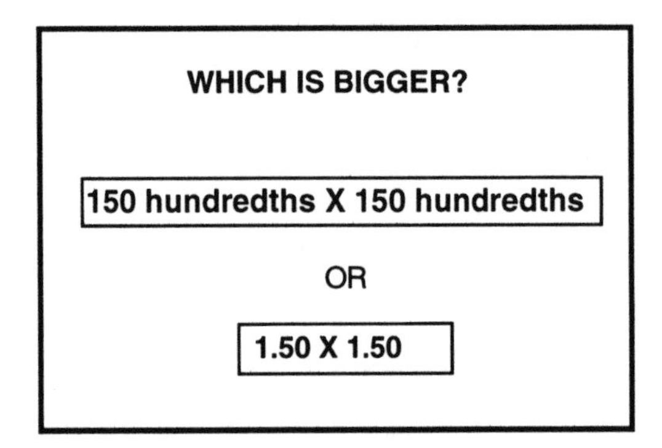

Jim picked 150 hundredths X 150 hundredths "because," he argued, "it's the same as 22,500 hundredths, or 225, whereas 1.50 X 1.50 is only 2.25." Was Jim right? Explain.

35 The Number Boxes

Using the numbers (4, 5, 7, 9) in the box, make as many fractions as you can that are less than one-half.

I am less than one-half.

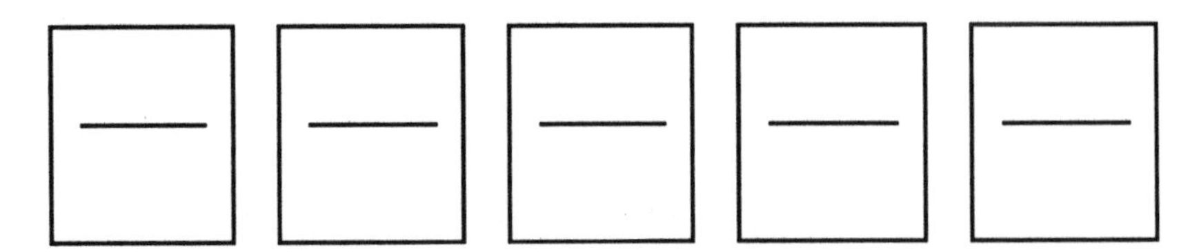

Using the numbers (.2, .01 .3, 4) in the box, make as many fractions as you can that are less than one-half.

I am less than one-half.

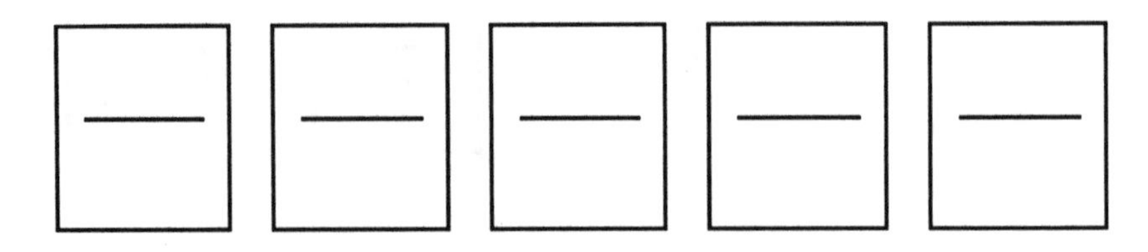

36 Not Enough Money

Alex went to the store to buy a gallon of ice cream that costs $3.60. He counted his 25 coins and found that he was a penny short. What coins did he have if only one of them was a nickel?

37 Your Turn to Explain

Sam divided 157 by 8. Here is his work:

$$\begin{array}{r} 19 \\ 8\overline{)157} \\ \underline{8} \\ 77 \\ \underline{72} \\ 5 \end{array}$$

Sam wrote the final answer as 19.5. What's wrong with Sam's answer?

38 Who's Taller?

Who is taller—Todd, who is 5.6 feet, or Scott, who is 5 feet 6 inches? Explain.

39 Switching Coins

Place 12 pennies in a row on the table. Replace every fourth coin with a nickel. Now replace every third coin with a quarter. Now replace every sixth coin with a dime. What is the value of the 12 coins on the table?

GEOMETRY

To the teacher

This section covers the concepts of congruency, area, perimeter and volume. The second question on Shadow's Pen assumes knowledge of circumference and area of circles. Some of the questions use the terms *equilateral*, *isosceles*, and *scalene*.

For Cutting Circles, encourage students to make a chart. Depending on your class, you can start the chart for them or let them develop their own.

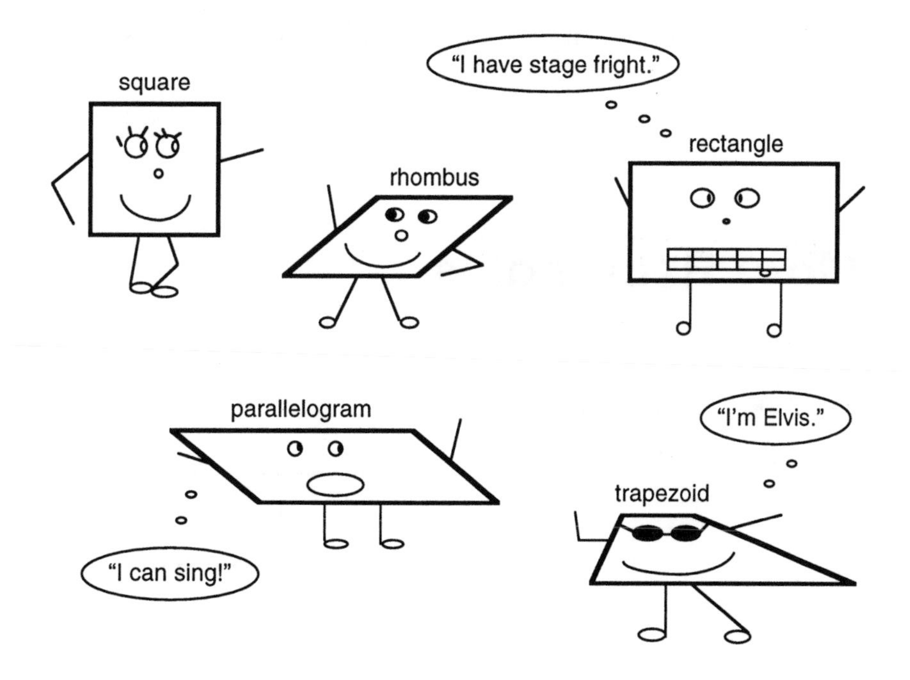

40 Counting Squares and Rectangles

Count the number of squares in the diagram below. (Don't forget overlapping squares!)

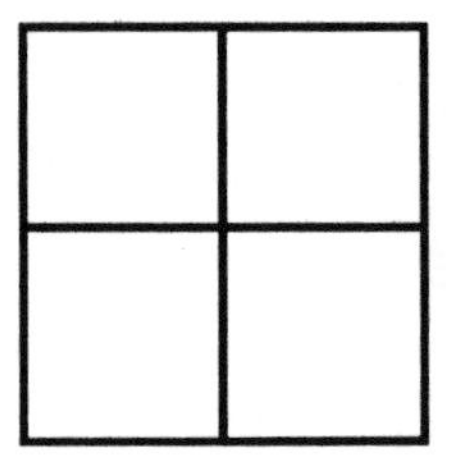

Now count the total number of rectangles. (Don't forget that a square is also a rectangle!)

41 More Rectangles

Count the number of rectangles below:

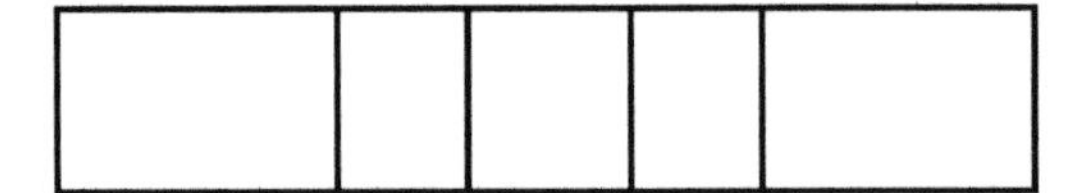

42 Counting Triangles

How many triangles can you find? (Don't forget overlapping triangles!)

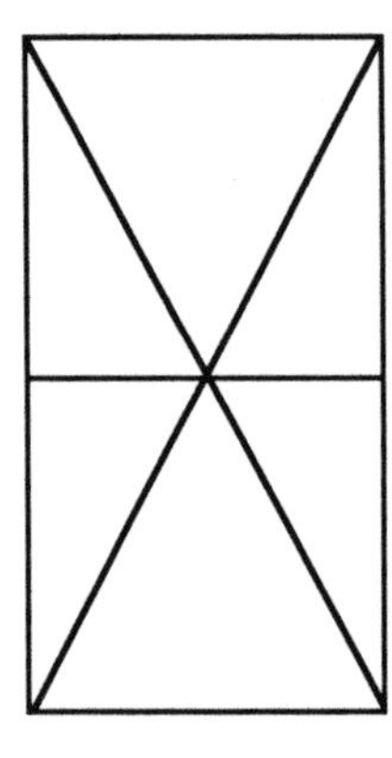

43 More Counting Triangles

How many triangles can you count? (Don't forget overlapping triangles!)

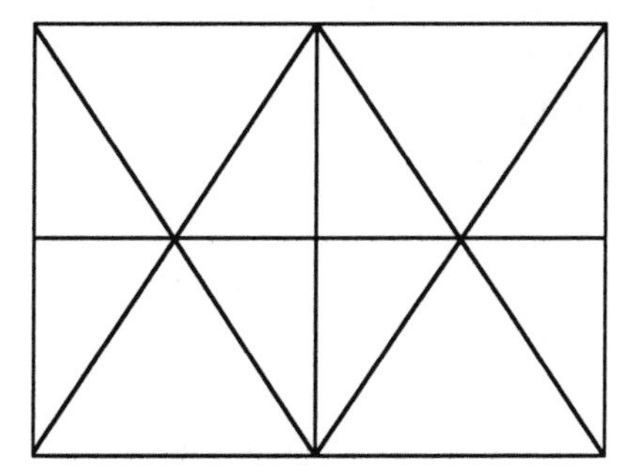

44 The Missing Perimeter

The diagonal of a square divides it into two triangles, each with an area of 18 sq. cm. (square centimeters). Can you find the perimeter of the square?

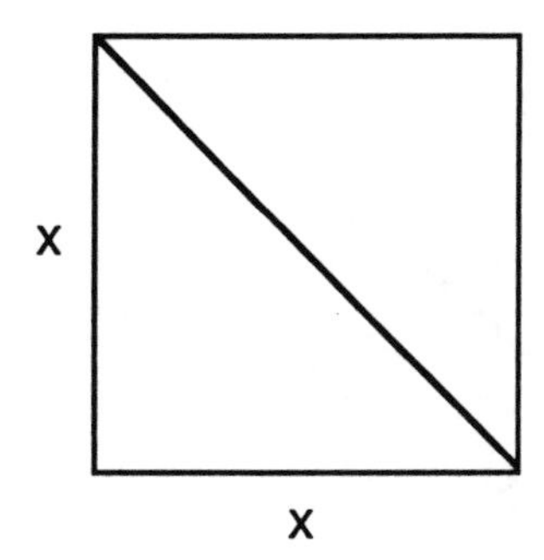

45 The Quadrilaterals Go Onstage!

The quadrilateral family has been invited to perform on stage. In order to get on stage, they must go through the doors below. To help the quadrilaterals find their way, list the correct quadrilateral names under each door.

Some quadrilaterals may be able to go through more than one door. Be ready to explain your answer.

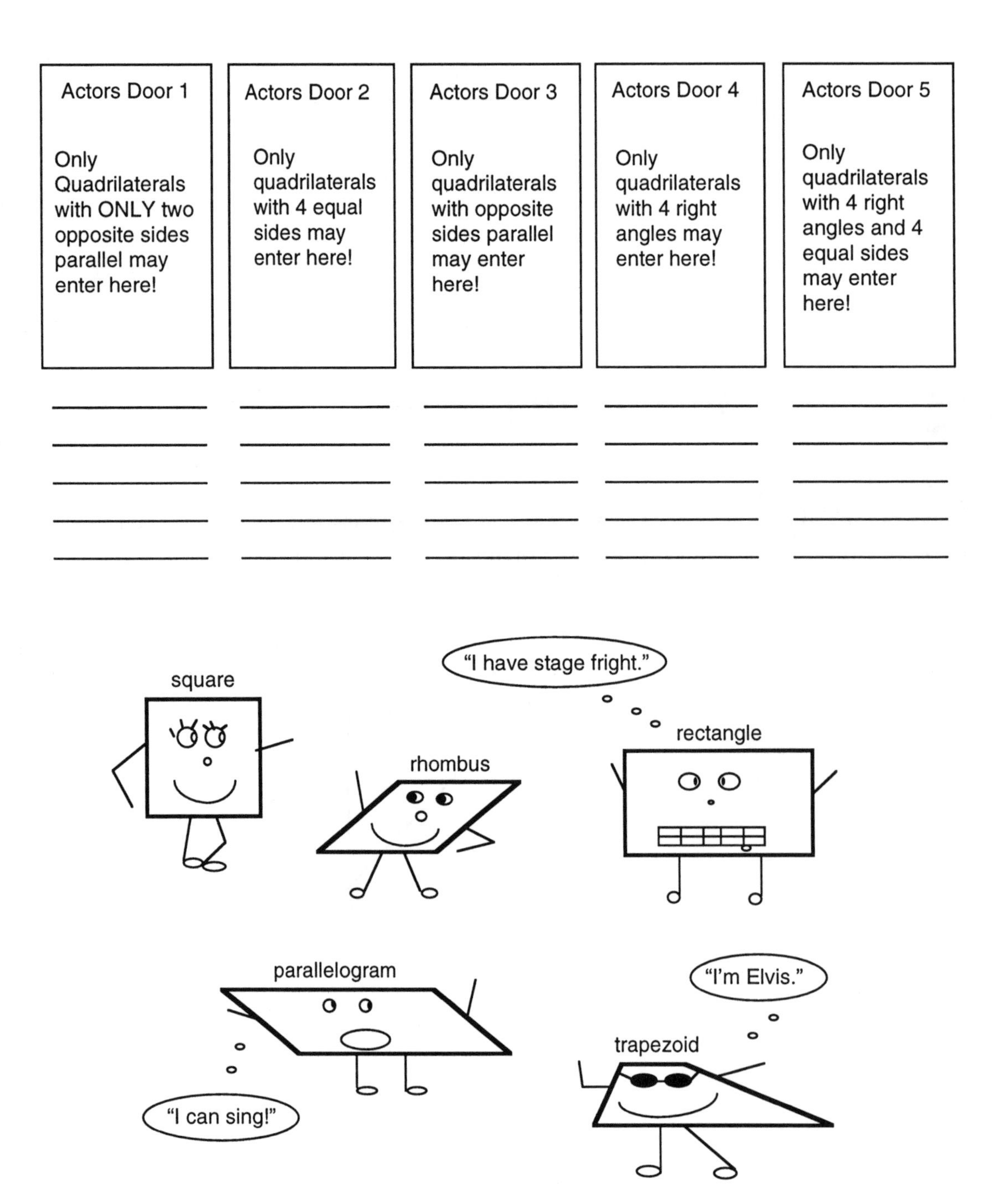

46 The Sum of Two Sides

The sum of two sides of a triangle must always equal *more* than the third side.
If y is a whole number, what values could it have?

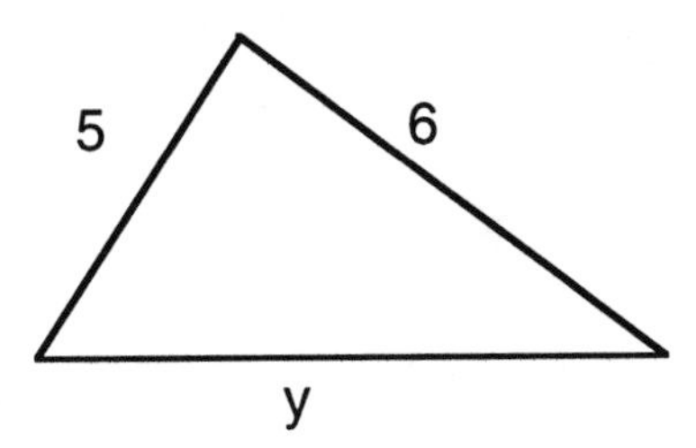

47 The Land Must Be Divided

Four daughters have to split this land into 4 equal parts. The parts must be congruent (they must have the same size and shape). Show how it can be done. Be sure to use pencil!

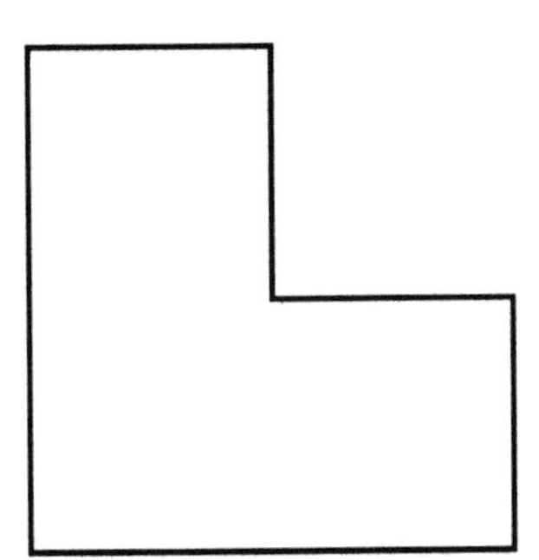

48 Types of Triangles

In the figure below, AB = BC. The sides that have the same markings are also congruent. Count and name the triangles that are isosceles, equilateral, and scalene. (Remember: every equilateral triangle is also isosceles.)

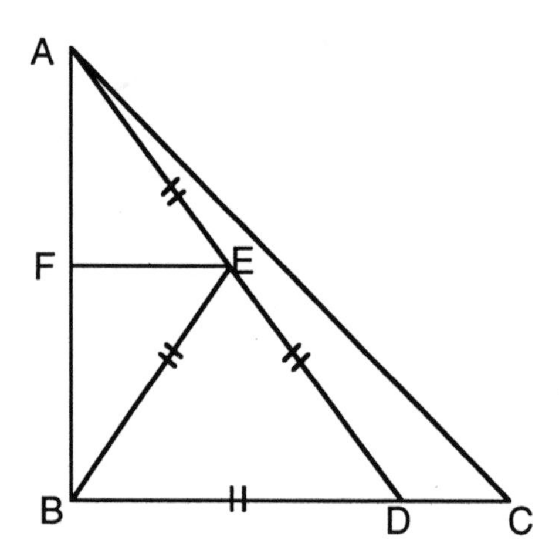

A. Number of equilateral triangles ______

Names of equilateral triangles ______________________________

B. Number of scalene triangles ______

Names of scalene triangles ______________________________

C. Number of isosceles triangles ______

Names of isosceles triangles ______________________________

49 Shadow's Pen

You want to build a pen for your new dog, Shadow, and you have 100 yards of fence. If you make the fence rectangular, what is the largest area you can enclose? (Make a chart.)

Now, let's say you want you want the pen to be circlular. How will the area for a circular pen compare to your answer above?

In Africa, many of the houses are shaped like circles. Native Americans also made their teepees round. Based on the results above, explain why this might be true.

50 Cutting Circles

Making only one cut, what is the largest number of sections you can make out of one circle? What is the largest number of sections you can make with two cuts? Three cuts? Four cuts? What about 8 cuts? (Use a pencil and the figures below to help you.)

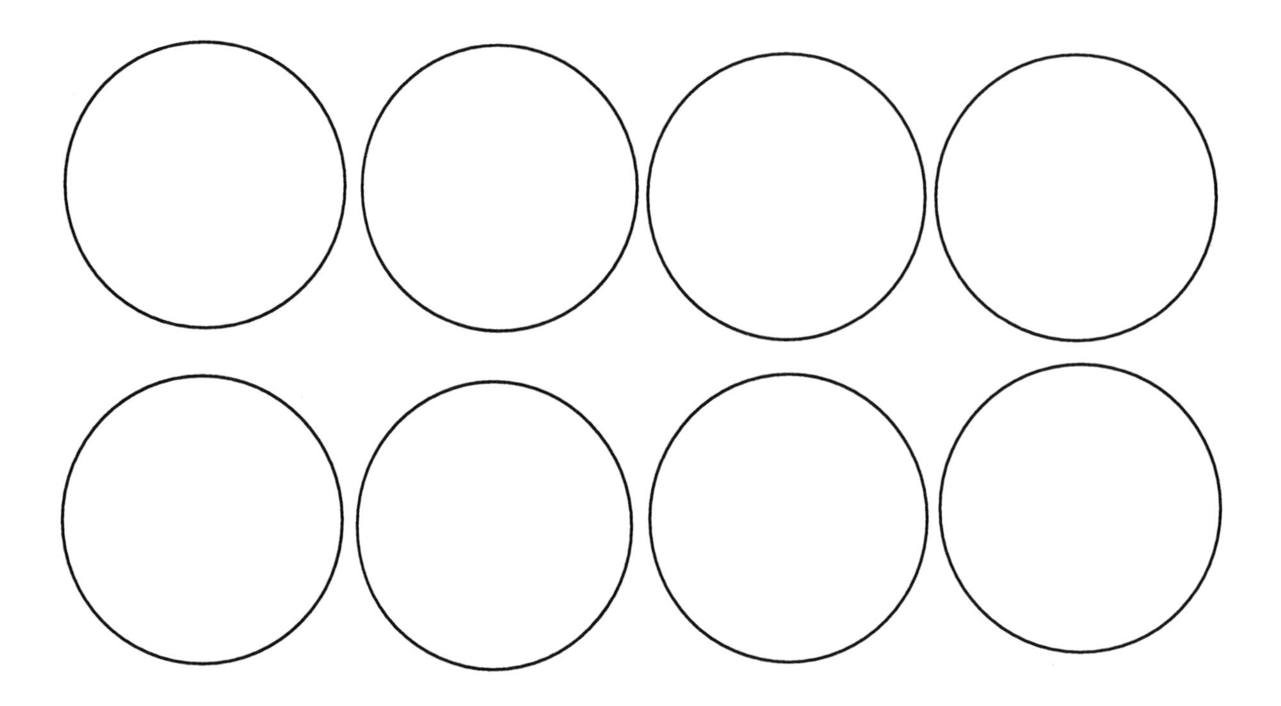

Can you find a pattern in the relationship between number of cuts and number of sections?

51 The Paint Fell on the Cube

A large can of red paint fell on this 3 X 3 cube, and all its surface, including its bottom, got covered with red paint.

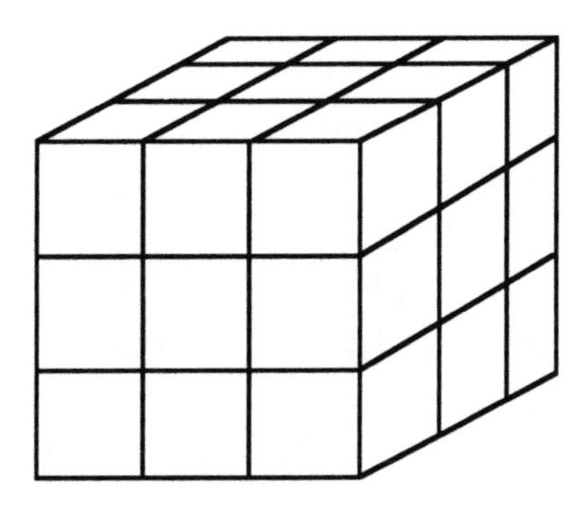

How many unit cubes have only one face painted red?

How many unit cubes have only two faces painted red?

How many unit cubes have three faces painted red?

How many unit cubes have no faces painted red?

RATIO, PROPORTION, AND PERCENT

To the teacher

This section includes ratio, proportion, and percents because students need to understand that percent is a ratio: the ratio of a number to 100. Also, I find that students gain more of an understanding of the concept of percent if I teach percent problems by using proportions. It is also important to bring decimals into any unit on percent, not only because it's easier to find tax amounts by using decimals, but also because students need to remember that *hundredths* means *percent*.

The ratio problems in this section are meant to be solved by using a chart; students can also use manipulatives.

In The Growing Baby, students need to understand how to change pounds to ounces. This problem provides an introduction to the concept of percent increase.

52 The Bird Shop

A bird shop at the mall sells only canaries and parrots. For every 5 birds sold, 3 are canaries and 2 are parrots. If the shop sells 45 birds in a week, how many of them are parrots? (Make a chart to help you.)

53 The Birthday Party

At a birthday party, there were five more girls than boys. If the ratio of girls to boys was 4 to 3, how many girls and boys were at the party? (Make a chart to help you.)

54 The Wedding Cake

If an 8″ x 8″ square cake serves 8 people, how many 24″ x 24″ square cakes are needed for a wedding attended by 144 people? (Make a diagram to help you visualize this problem.)

55 Percent City

In Percent City, all the buildings must be 100 floors. In Five Town, all the buildings are as tall as the buildings in Percent City, but they have only five floors. This allows for storing very large machinery.

If you are in an elevator halfway between floors 4 and 5 in Five Town, what floor would that be in a Percent City building? Finish labeling the floors to help you find the answer. (Level 0 is the ground floor.)

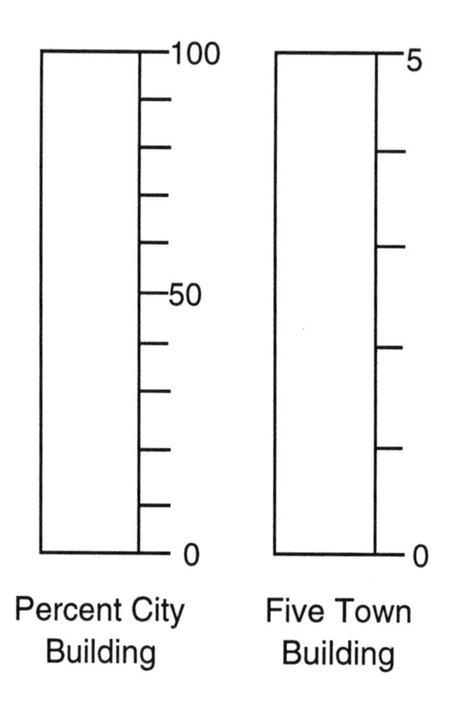

56 Is It Free?

A computer game used to cost $50.00. Then it sold at a 50% discount. Sandra decided to wait because she didn't have enough money. During the next week, the price again went down 50%. Jim, Sandra's younger brother, thinks that Sandra can now get the computer game for free. He explained that 50% plus 50% is 100%. So if it's discounted 100% then it's free. Do you agree with Jim?

Explain.

57 The Growing Baby

Sarah's baby brother weighed 8 lbs. when he was born. In a week, his weight decreased (dropped) by 2 lbs. 4 ounces. A month later, the doctor said he had gained 50% of his last weight. How much does the baby weigh now?

58 The Sales Tax

Emma went out shopping with her father and bought a dress that cost $40.00. In class, she learned to find the tax by multiplying by .08 (the sales tax in her state is 8%). Emma found the tax and then added the tax to the original amount. Emma's father suggested that she should just multiply the cost of the dress by 1.08. Emma was confused. Who is right? Work it out both ways and explain your results.

PROBABILITY

To the teacher

Probability is best taught by doing experiments and playing games, but it is important to take time to discuss results. Students will often generalize results and mistakenly apply them to different kinds of problems. Problems 59 and 60 (dice-rolling problems) can be used to demonstrate how similar problems can generate very different results (the two problems should be done together).

The Cayuga Indians game can be made easily with large lima beans colored with a black mark on one side. The game encourages students to think about why certain toss combinations (5 or 6 out of 6) are more likely than others (3 or 4 out of 6). A discussion of details may be too advanced at this level, but students should realize which kinds of combinations are harder to get.

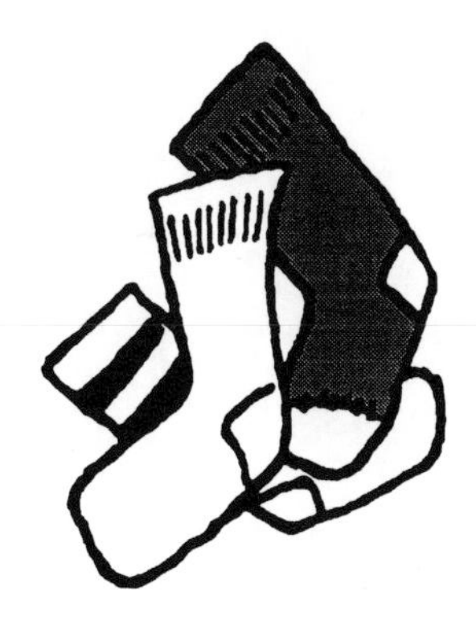

59 Roll One Die

Roll one die at least twenty times (the more rolls, the better) and diagram your results as shown below (two dots over the 2 means the die showed two dots on two diferent rolls).

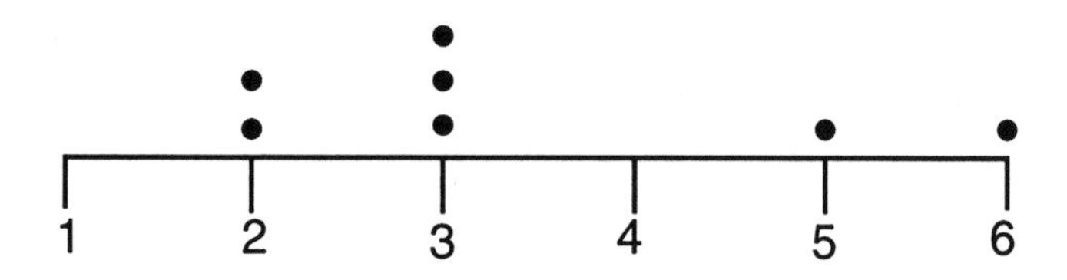

A new number line is provided below.

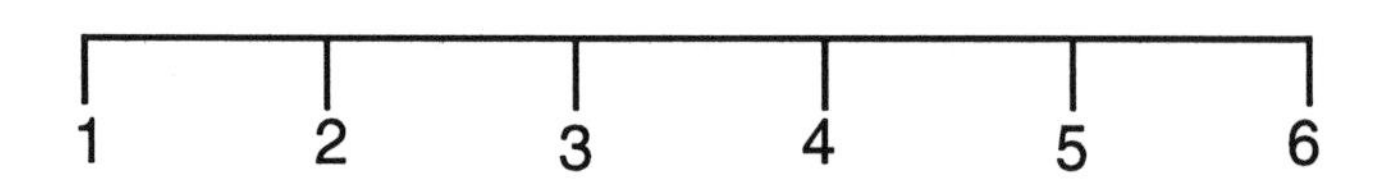

How many times did you get 1? ______ 2? ______ 3? ______ 4? ______
5? ______ 6? ______

For each number below, write a fraction to show the number of times you got that number out of the total number of times you rolled the die (reduce your fraction if possible).

1 ________ 2 ________ 3 ________

4 ________ 5 ________ 6 ________

Summarize your results:

Roll Two Dice

Roll two dice at least twenty times (the more rolls, the better). What sums could you get? Circle the *sums* below that are possible:

1 2 3 4 5 6 7 8 9 10 11 12 13 14 15

Keep a tally of the number of times you get these sums, then graph your results:

Sums	Tally

Number of Tallies

2 3 4 5 6 7 8 9 10 11 12

Sums

Explain your results. Compare these results to the results of rolling only one die.

61 Let's Sit Together!

Nadia, Tobi, and Elena have gone to their school play and want to sit together. In how many different ways can they sit together? Using the diagram below, finish finding all the possible seating arrangements. (Two arrangements are shown.)

1. Nadia Tobi Elena
2. Nadia Elena Tobi

62 Amanda's Name

At Amanda's birthday party, she puts each letter of her name into a bag. Each of her letters has a point value, as shown below:

A = 1 point

M = 5 points

N = 5 points

D = 5 points

Everyone at the party must reach into the bag five times. Each time, they must get a letter (without looking), record the points for the letter, and then put the letter back in the bag. At the end of five tries, Mark has 9 points. How can this be?

Which letter is more likely to come out each time?

63 The Socks Problem

Hanna's stepfather is a volunteer fireman. He wears only black socks and white socks. One night after a storm, the lights go out and he must go into town to help put out a fire. He gets up in the dark, reaches into his drawer, and grabs one sock. How many more socks must he grab to make sure he has a matching pair?

64 The Cayuga Indians

The Cayuga Indians in upstate New York played many probability games. In one of their games, they used 6 flattened wild plum seeds, blackened on one side and plain on the other side. There were only four winning throws, as shown in the top two and bottom two rows below:

●	●	●	●	●	●	4 points
●	●	●	●	●	○	1 point
●	●	●	●	○	○	0 points
●	●	●	○	○	○	0 points
●	●	○	○	○	○	0 points
●	○	○	○	○	○	1 point
○	○	○	○	○	○	4 points

If you made a throw and got no points, you would lose your turn. Explain why you think the Cayuga Indians decided on the point system above.

Play this game with a friend. How hard is it to get 4 points?

SETS

To the teacher

Venn diagrams are a great strategy for solving certain kinds of problems, including all the ones in this section. Students will need to see some problems solved with this method before they know how to proceed on their own (The Family Reunion is a good problem to demonstrate). Once students understand the procedure for using Venn diagrams, they enjoy making their own problems for others in the class to solve.

For the problems in this section, you can use manipulatives such as little cubes or circle rings; you can even create your own.

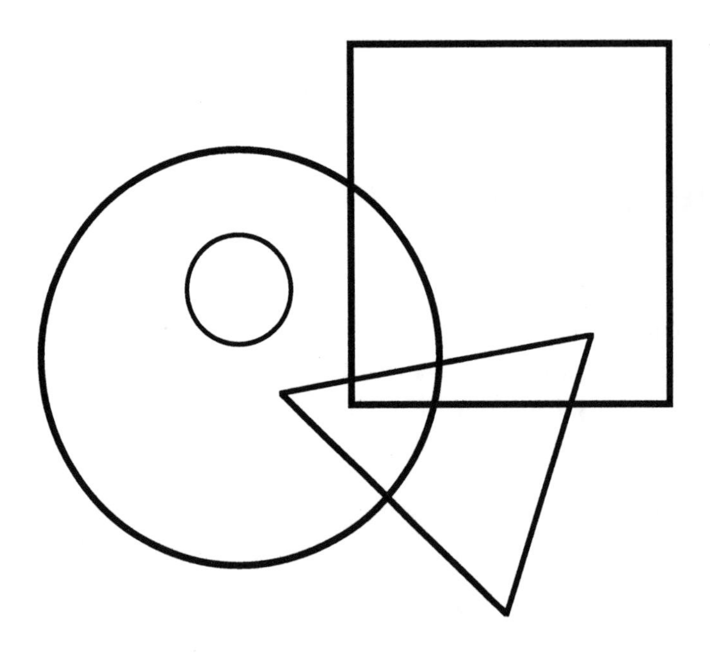

65 The Family Reunion

The Smith family and the Perez family were having a joint family reunion. The Smith family was expecting ten family members, and the Perez family was expecting fifteen. Of those totals, two Smith brothers had married Perez sisters. All four were considered members of both families. How many people were at the reunion?

Finish the Venn diagram to help you.

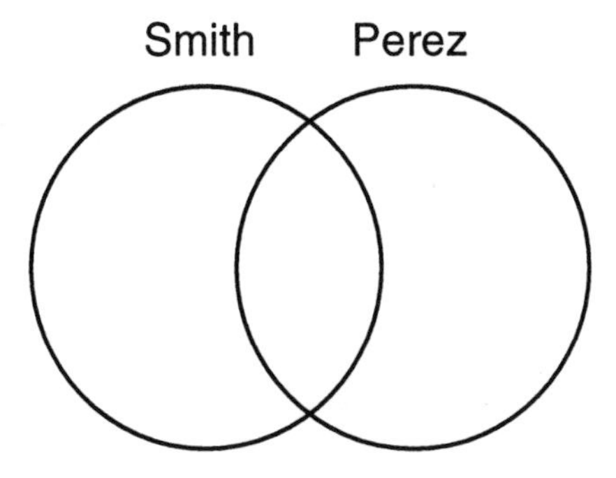

66 Pine Elementary School

At Pine Elementary School, there are three clubs. The music club has 20 students, the games club has 50 students, and the computer club has 25 students. Five students in the music club are also in the computer club but not the games club. One student is in both the computer club and the games club but not the music club. Two students are in the music club and the games club but not the computer club. Three students are in all three clubs. How many students participate in clubs?

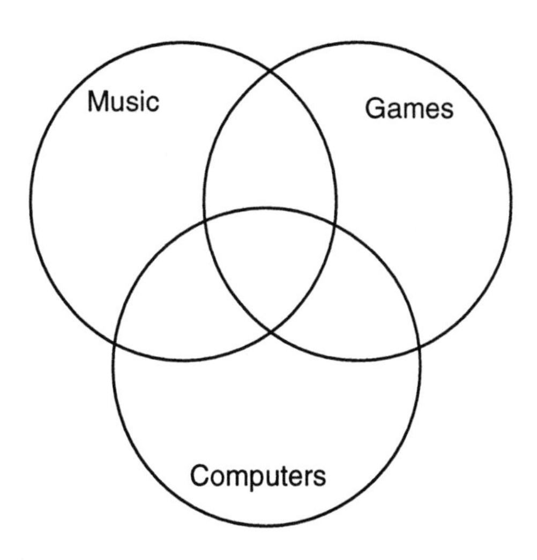

The Marble Collection

Scott called Lucia on the phone because he wanted to buy her marble collection. "How many marbles do you have?" asked Scott. Lucia responded, "I know I have some very old marbles. Ten of those are red. Six are red and new. I know I have 20 new marbles. Four of the new ones are blue. Three of the new ones are brown."

Scott came prepared to buy 43 marbles. Is that how many marbles Lucia really has?

Make Your Own Sets Problem

____

____

____

Solve it, using a diagram:

PRE-ALGEBRA

To the teacher

The following problems were provided to help students start thinking about algebra concepts such as balancing an equation, inverse operations, and using a variable to represent a number.

The problem Three Times My Age? is a good one to also include in the ratio and proportion unit.

Allowing students to experiment with algebra concepts without formally teaching algebra will help students be stronger in algebra when they encounter it in middle school and later. The more you expand your topics at this level and allow students to think deeper and more critically, the more you help them build a good foundation for mathematics.

Mr. Function Machine

69 A Calculator Mistake

Using a calculator, a student accidentally multiplied by 10 when he should have divided by 10. The incorrect answer displayed was 300. What is the correct answer?

70 Another Calculator Mistake

Using a calculator, a student accidentally divided by 60 when he should have multiplied by 60. The incorrect answer displayed was 10. What is the correct answer?

71 What values?

For what *values* of x is $x + x = x \bullet x$? Explain your answer.

For what *values* of x is $x + x + x = 3 \bullet x$?

72 Three Times My Age?

Sally turned to her mom and said, "Mom, you are 35 and Jimmy is 5, so you are seven times his age. Will you always be seven times his age?" Mom answered, "No, someday, I'll be four times his age. As times goes by, I'll be three times his age, then someday I hope to be twice his age, etc." Sally was confused. Can you explain why Mom was right? Make a chart to help you.

73 Consecutive Whole Numbers

Consecutive whole numbers are whole numbers in a row, like 5, 6, 7 or 12, 13, 14, 15.

Can you find three consecutive whole numbers whose sum is 375?

74 Balance It!

The bag on the left side of the scale below contains a number of blocks. All blocks on the scale are of the same size and weight and, as you can see, the scale balances perfectly. Explain how you would find out the number of blocks in the bag if you couldn't peek.

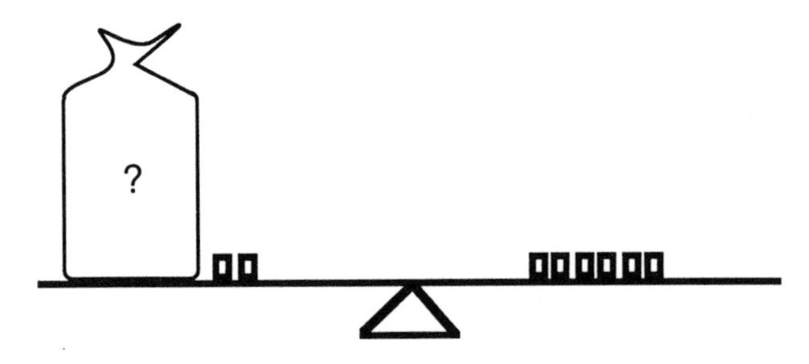

75 Balance It Again!

Each bag has the same number of blocks inside, and all the blocks on the balance scale are the same size and weight. As you can see, the scale balances perfectly. Explain how you would find out how many blocks are in each bag if you couldn't peek.

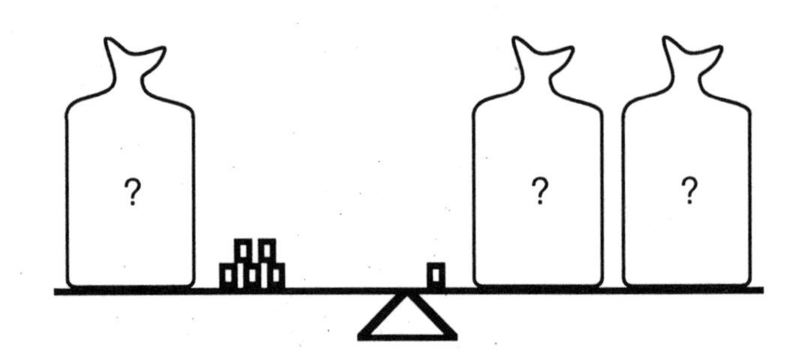

76 Make Your Own Story

Make your own story problem that can be illustrated by this number sentence:

$$\square - 3 = 10$$

77 The Function Machines Are Running Wild!

Sarah, the company's chief engineer, walks into the function machine room and finds all the machines running. Her job is to figure out what each machine is doing before turning them all off. Can you help her figure this out?

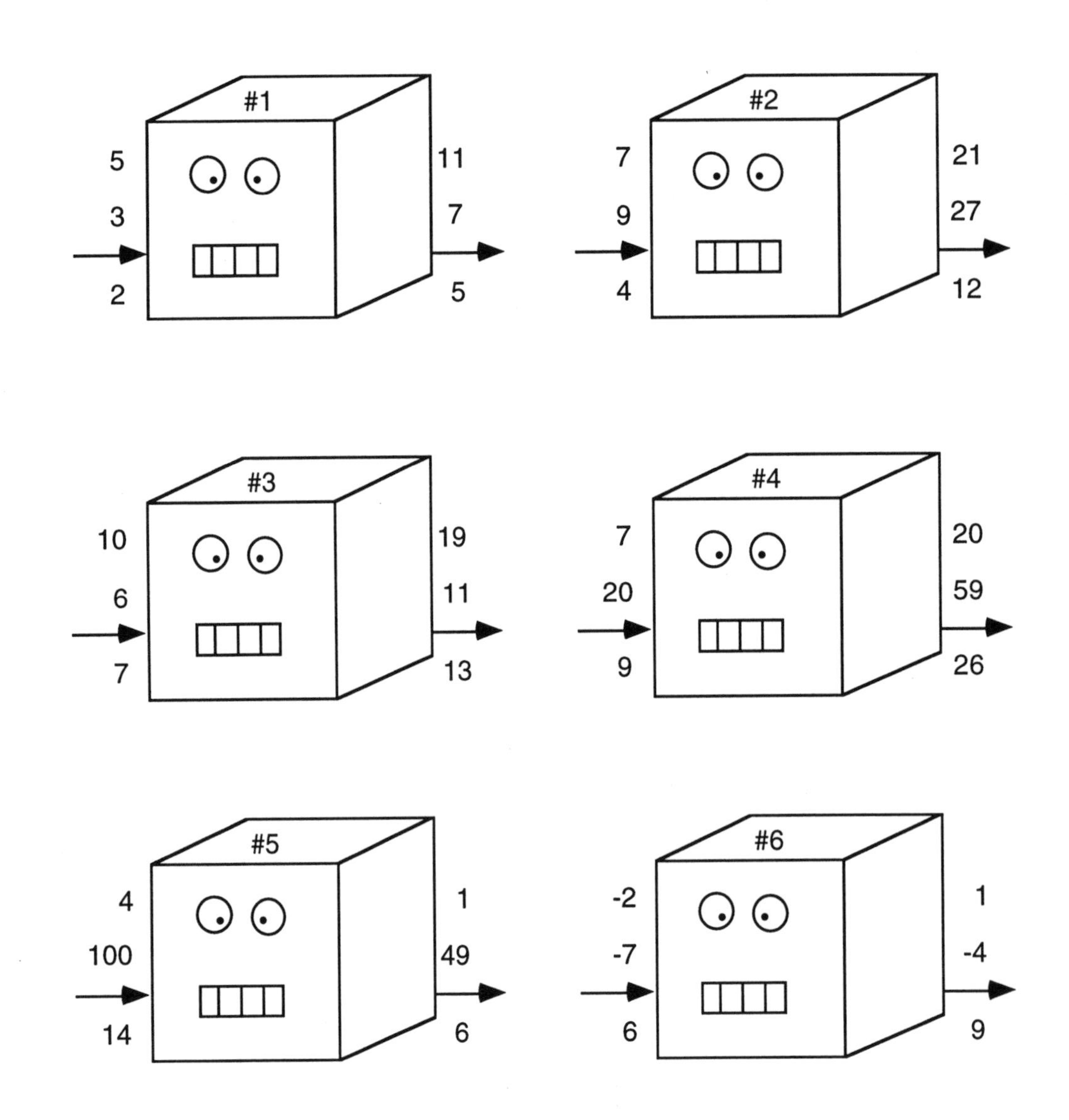

Now, can you make your own function machine?

MORE CHART AND POPULAR PROBLEMS

To the teacher

This section includes more problems that require making a diagram or chart. I've also included some problems that are especially big hits with upper elementary students. Can You Walk Through Paper? shows students that we often get stuck on only one way of thinking and that looking at things differently may solve the problem. Another favorite is A Coloring Problem. For more "topology" problems, see *Math Ties Book B1.*

For this section, you may need coloring pencils or crayons, scissors, and pencils with erasers.

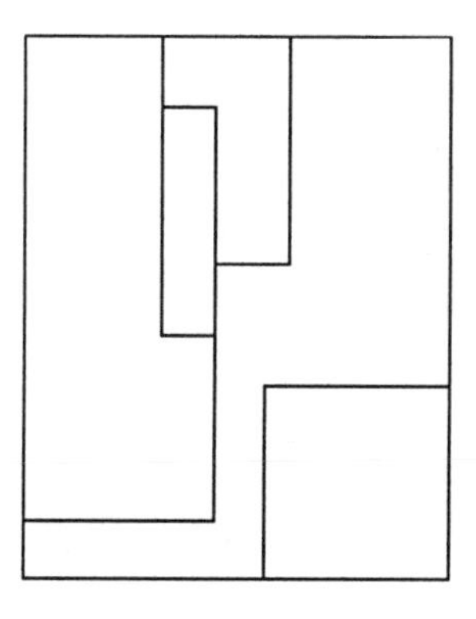

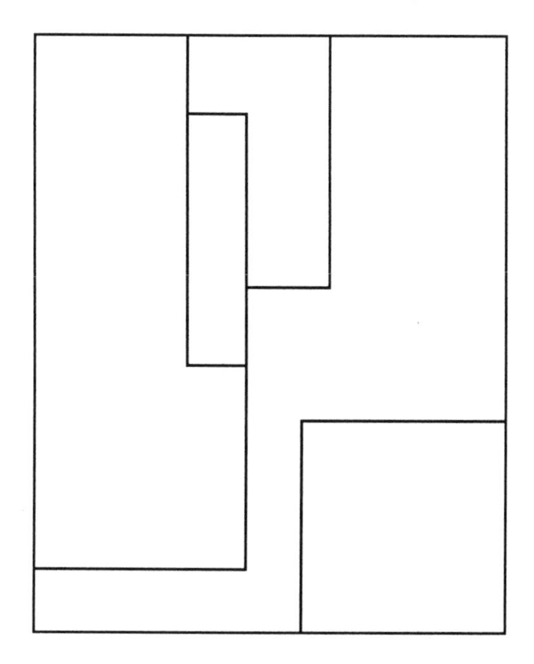

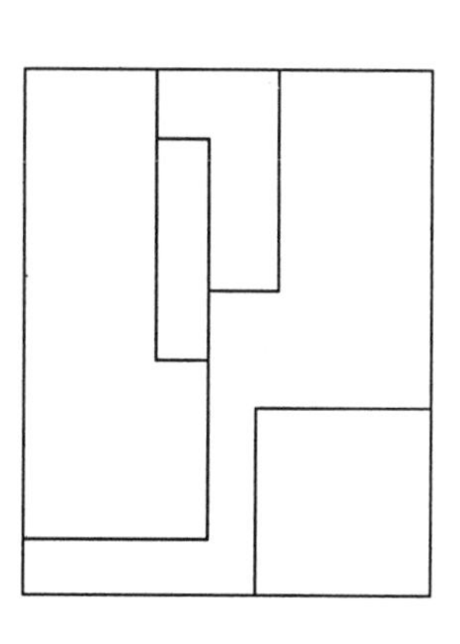

78 Cutting Up

A piece of fabric is 6 yards long. If you want to cut it into 1-yard pieces, how many cuts do you have to make? (Try it before you decide on an answer!)

79 The Clock Strikes Again

When a clock strikes 6, it takes five seconds between the first and sixth strike. How long would it take the same clock at midnight between the first and twelfth strike?

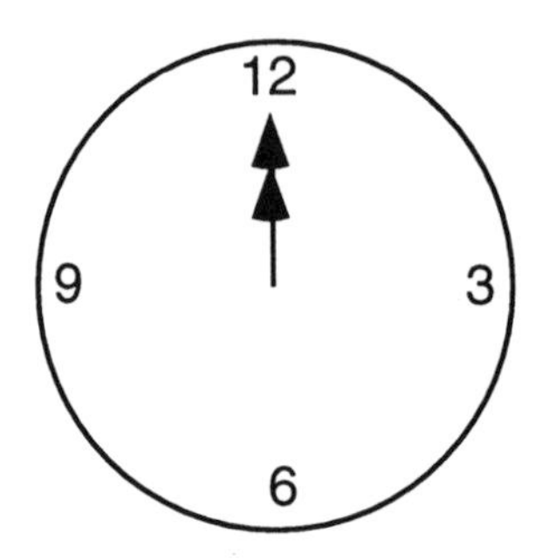

80 Let's Sit In A Circle

If fourteen people sit, evenly spaced, in a circle to hear a story, who sits diagonally across from person 3?

The Highway

A highway is 78 feet wide. On each side, it has a shoulder that is 12 feet wide. It has a median divider that is 3 feet wide. How wide is each travel lane of the highway if the road is divided into 4 lanes of equal width (two going one way and two the other)? Give your answer in feet and inches.

82 A Sign For Your Name

A sign is 35 inches wide. You want your first name to be centered on the sign. You must use block letters each 1 inch in width, leaving a 1/2 inch space between letters. How far from the left-hand side of the sign is the first letter of your name?

83 Visiting Your Friend

In how many ways can you get from your house to your friend's house? (Use the diagram below, and follow the arrows.)

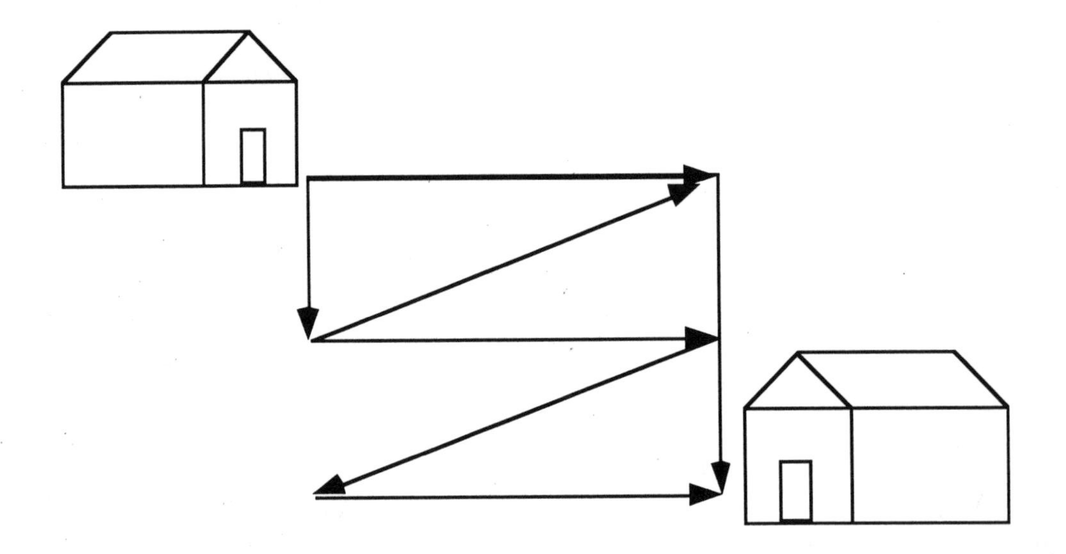

84 Can You Walk Through Paper?

Can you make a hole on a regular piece of paper and walk through that hole? Give up? Take a regular sheet of paper and draw this:

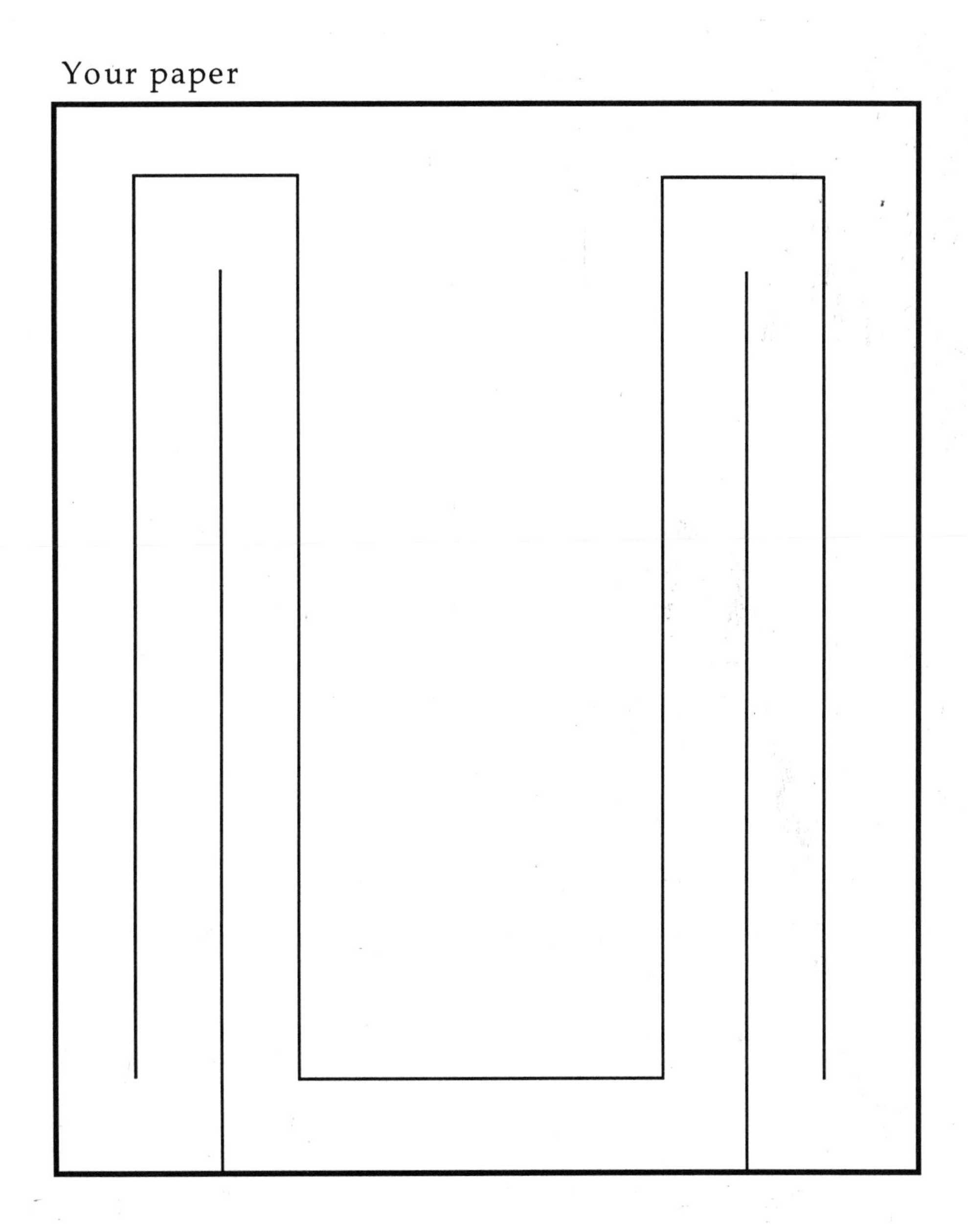

Now that you've drawn this on your own paper, cut carefully along the lines you've drawn. See if you can walk through the opening!

85 A Coloring Problem

Try to color this map with the least number of colors possible. No areas side by side may have the same color, but two areas of the same color may touch at a point. What is the least number of colors you can use?

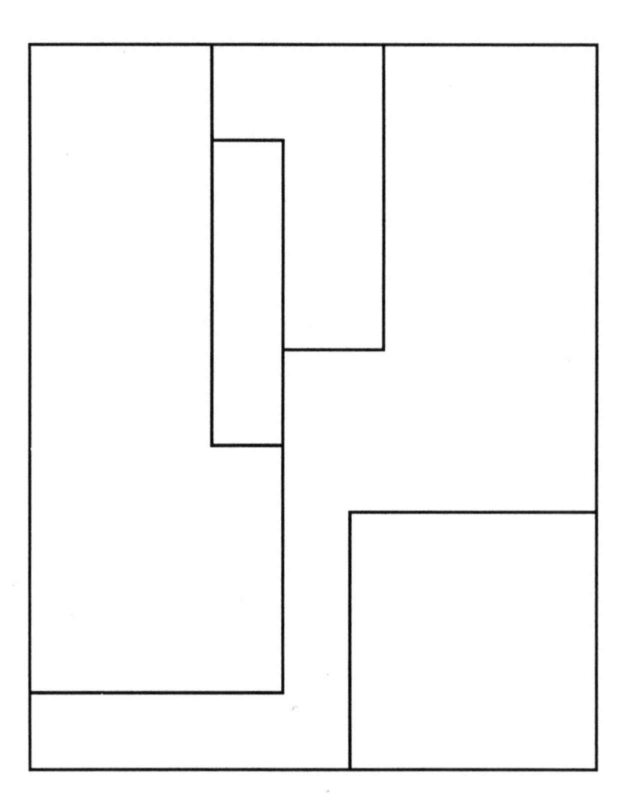

Now, can you create a map where you need more colors?

Answers

LOGIC PROBLEMS

1. The Chicken, the Corn and the Fox

The man takes the chicken to the other side (he leaves the fox with the corn since the fox does not eat corn). He returns, picks up the corn, and takes it to the other side. He brings the chicken back, drops it off, and picks up the fox. He leaves the fox on the other side with the corn. Then he returns to get the chicken.

2. A Family Problem

Here is one solution: The husband crosses the river with the daughter (160 + 85 < 250). The daughter returns and stays. The wife and son cross together (130 + 115 < 250). The wife returns. Finally, the wife and daughter cross the river (130+ 85 <250).

If you cannot leave the daughter alone at any time, here is another solution:

The mother and son cross the river (130 + 115 < 250). The son returns. The father and daughter cross the river (160 + 85 < 250). Then the mother returns to pick up the son, and both cross the river.

3. Truthful?

If the student never tells the truth, the statement "I never tell the truth" must be false, so there is a contradiction.

4. The Liars Club

The students finished the spelling bee in this order: 1-Tom, 2-Katie, 3-Alex, and 4-Susie.

Since all statements are false, we can fill in a chart by crossing out the "false" boxes. (We know Katie didn't finish first and Alex didn't come in last, nor did he beat Katie, so Alex is third while Katie is second. Susie didn't beat Tom, so she's last, and Tom is first.)

	1	2	3	4
K	X	O	X	X
A	X	X	O	X
S	X	X	X	O
T	O	X	X	X

5. Liza at the Pond

Liza fills the 5-cup container completely, then pours it into the 3-cup container (leaving 2 cups in the 5-cup container). She pours the 2 cups into the bucket and then repeats the process to get 4 cups.

6. Line Up!

The friends from youngest to oldest are Peter, Juan, Tammy, and Carmen.

The friends from shortest to tallest are Juan, Peter, Carmen, and Tammy.

7. The Blocks Puzzle

Here are the figures that are opposite each other:

trapezoid—parallel lines

rectangle—circle

triangle—pentagon

WHOLE NUMBERS

8. The Missing Symbols

Inside the parentheses, multiplication is done before addition (by order of operations).

(1 • 2 + 3) 4 • 5 = 100

9. The Missing Steps

From the first floor to the second floor are 26 steps, so we can assume there are 26 steps between floors. From floor 1 to floor 6 there must be five sets of 26 steps, or 130 steps.

10. The Cool Company

Here is one way to meet the customer's request: three bags of 5 pounds each, 2 bags of 12 pounds each, 1 bag of 20 pounds each, and 2 bags of 35 pounds each.

11. The Slippery Well

The frog gets out on the 17th day. He hits the top of the well (0 mark) on the 16th day, but is not out until the next day. Once he's out, he should stop slipping down!

12. The School Math Competition

The student receiving 34 points got eight problems right and two problems wrong. (Eight right, or 40 points, minus two wrong, or 6 points, is 34 points.)

13. The Toy Factory

It would take 14 1/2 hours to make 5 bikes and 10 dolls ($5 \times 1.5 = 7.5$; 10×42 min. = 420 min. or 7 hr.; $7 + 7.5 = 14.5$). If they started at 6:30 A.M., they would be done at 9 P.M.

14. Todd's Turtle

Todd made $12.00 (–$15 + $20 – $23 + $30 = $12).

15. The Hockey Camp

Plan A is better if all the brothers go; they save $30 (they pay $400, compared to $250 + $60 × 3 = $430). If Brian doesn't go, Plan B is better, and they save $30 (they pay $250 + $60 × 2 = $370, compared to $400).

NUMBER THEORY

16. The Never-ending Joke

After 5 minutes, 127 friends know the joke. ($1 + 2 + 4 + 8 + 16 + 32 + 64 = 127$)

17. The Presidents

With 5 presidents, there would be a total of 10 handshakes. The following chart illustrates how (each person is represented by a number):

1 & 2	2 & 3	3 & 4	4 & 5
1 & 3	2 & 4	3 & 5	
1 & 4	2 & 5		
1 & 5			

If there were 10 presidents at the meeting, there would be 45 handshakes:

1&2	2&3	3&4	4&5	5&6	6&7	7&8	8&9	9&10
1&3	2&4	3&5	4&6	5&7	6&8	7&9	8&10	
1&4	2&5	3&6	4&7	5&8	6&9	7&10		
1&5	2&6	3&7	4&8	5&9	6&10			
1&6	2&7	3&8	4&9	5&10				
1&7	2&8	3&9	4&10					
1&8	2&9	3&10						
1&9	2&10							
1&10								

18. Mail in Remote Land

Groceries, mail, and clothes arrive on the same day every 60 days. (Find the Least Common Multiple of 3, 4 and 5.)

19. TV Cartoons

They both start again in 180 minutes (least common multiple of 90 and 60) or in three hours, which will be at 10 A.M.

20. The Salary Problem

1 cent a day, 2 cents a day, etc. is much better!

1 + 2 + 4 + 8 + 16 + 32 + 64 + 128 + 256 + 512 + 1024 + 2048 + 4096 + ...+262144 (19th day) +...

Note that the total for each day has to be added to the totals from all previous days to obtain the final answer—quite the task!

21. The Narrow Staircase

Only 40 more bricks will be needed to make a total of 55 bricks for a staircase that has ten steps.

22. The Proud, Perfect Squares

The next perfect square is 25.

The next two perfect squares are 36 and 49.

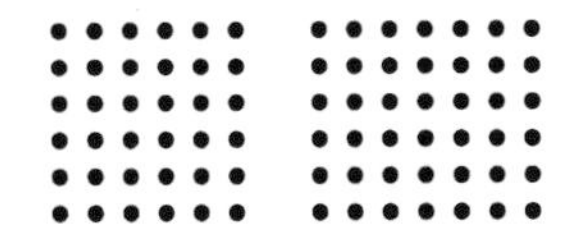

23. The Famous Odd Numbers

The sums are 4, 9, 16, 25, 36, then 49, 64, 81, 100 (the perfect squares up to 100).

The number of digits being added is squared to give the answer ($1 + 3 + 5 = 3^2 = 9$).

24. The Bag of Marbles

You have 35 marbles. Start by using a hundreds chart to make it easy to eliminate multiples. When you count marbles and get a remainder, realize that the number of marbles is not divisible by the counting number, so it is not a multiple of the counting number. Since counting by 2 leaves a remainder, you can eliminate all multiples of 2. However, it's easier to first eliminate all multiples of 10 since you know 10 is a multiple of 2. When divided by 5, the number has no remainder, so multiples of five should be explored to see which works. (You could start with multiples of 5, but a multiple of 5 may also be divisible by 10. In eliminating the multiples of 2, you get numbers that end with 5, but not numbers ending with 0).

25. Going to the Beach

There were 62 going to the beach:

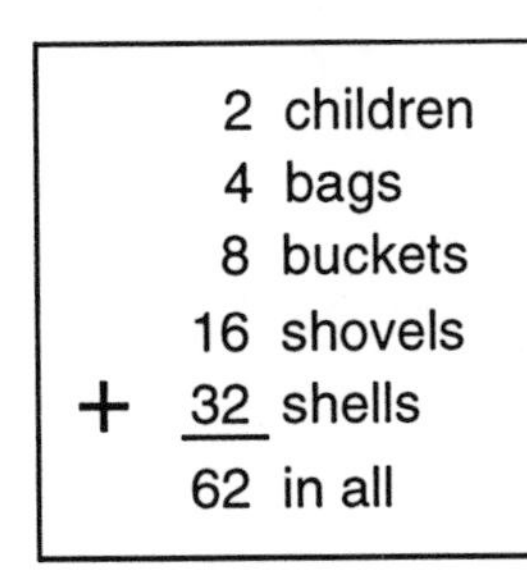

FRACTIONS

26. Fraction Diagram

The diagram below illustrates 2/3 of 9. The answer can also be shown with blocks, drawings of stick people, etc.

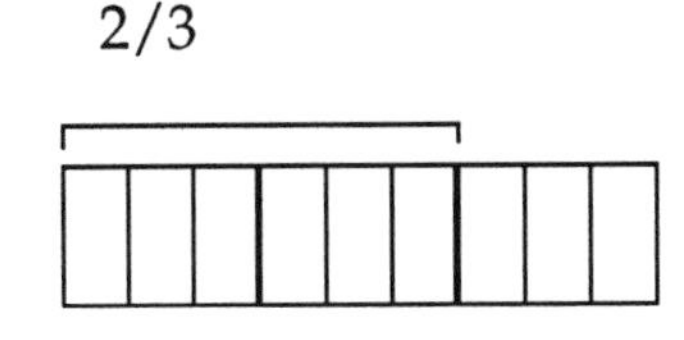

27. Diagram of Difference

The following diagrams show the difference between $10 \div 2$ and $10 \div 1/2$.

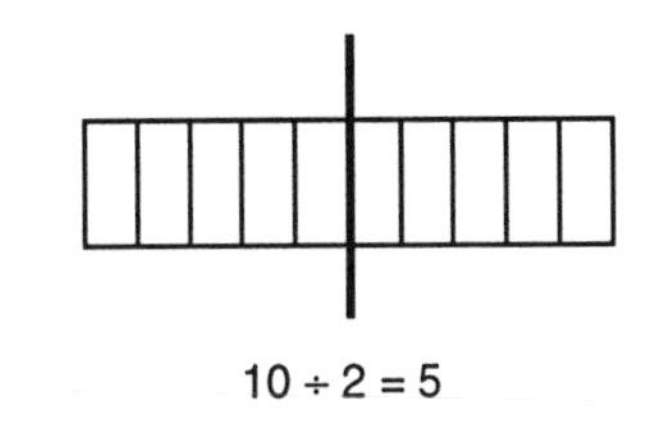

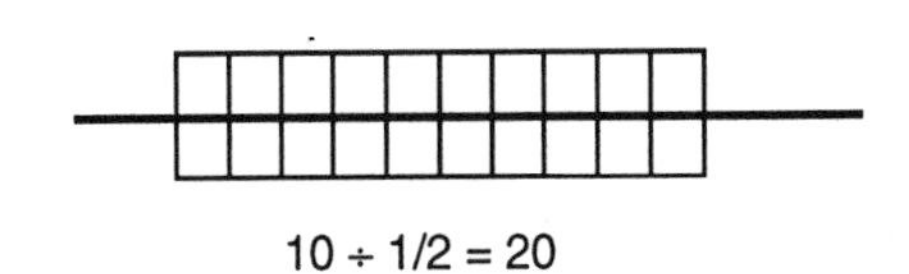

To find $10 \div 1/2$, ask How many halves are in 10? $10 \div 1/2 = 20$ (Remember that $10/1 \div 1/2 = 10/1 \times 2/1$.)

28. Wanted!

Answers may vary.

For a fraction to be larger than 1/8 (if you keep the numerator of 1) the denominator has to be *smaller* than 8; for a fraction to be smaller than 1/4 (if you keep the numerator of 1) the denominator has to be *larger* than 4. Because they are each smaller than 1/4 but larger than 1/8, the fractions 1/5, 1/6, and 1/7 are possible answers.

29. The Mystery Fraction

The missing fraction is 12/1, or 12. The

given fraction problem is the same as 1/2 × 2/3 × 1/2 × 1/2 × [empty box] = 1.

30. Ms. Kaplan's Musicals

The musical lasts 2 hours and 8 minutes.

If it ends at 10:30, the latest time the musical could start is 8:22 P.M.

31. The Bouncing Ball

The ball rises 3 1/8 feet (3.125 ft.), or 3 feet 1 1/5 inches. When it hits the ground at bounce 6, the ball has travelled a total of 293 feet 9 inches.

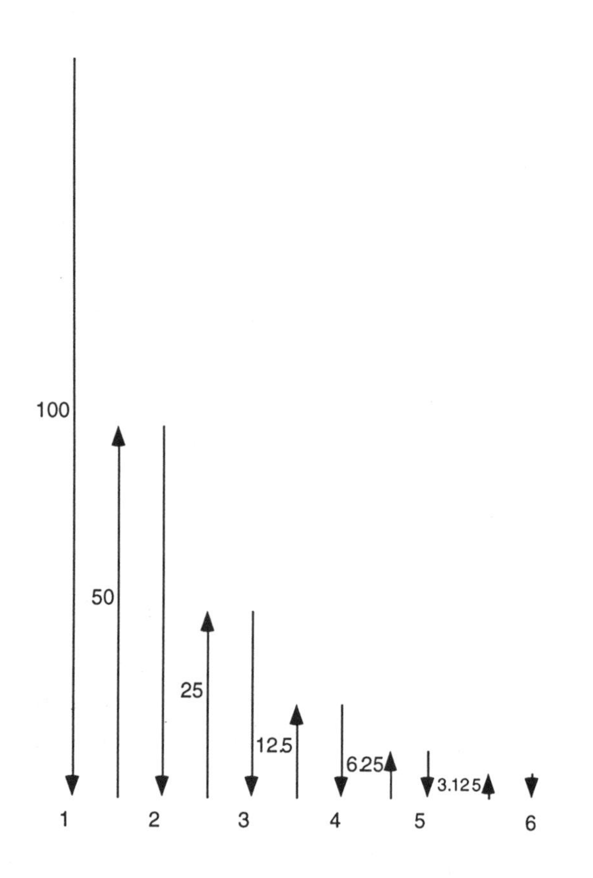

32. The Tricky Fence

Each space is 7 1/3 inches. Between the two end posts (outer edge to outer edge), there is 8 feet, or 96 inches. Part of the 8 feet is taken up by the 10 posts at 3 inches each (30 inches, or 2 1/2 feet). Subtract 96 – 30 to get 66 inches, or 5 1/2 feet. Divide 66 by the 9 spaces between posts to get 7 1/3 inches per space.

DECIMALS

33. Place Value Game

Answers will vary, depending on luck! Of course, to get the greatest possible number, each player should write the drawn numbers highest to lowest.

34. Which is Bigger?

No, Jim was wrong. Students may be likely to multiply 150 hundredths times 150 hundredths and not consider the decimal point. 150 hundredths is the same as 150/100, so 150/100 × 150/100 is 22,500/10,000, which gives the same answer as 1.50 × 1.50.

35. The Number Boxes

If you use 4, 7, 9, and 5, only 4/9 is less than one-half.

If you use .2, .01, .3, and 4, the following fractions are less than one-half: .2/4, .3/4, .01/4, .01/.2, and .01/3.

36. Not Enough Money

Alex had one nickel, four pennies, ten dimes, and ten quarters.

37. Your Turn to Explain

The correct answer is 19 5/8, or 19.625, (not 19.5).

38. Who's Taller?

Todd is 5.6 feet tall, and Scott is 5.5 feet tall (since he is 5 feet 6 inches), so Todd is taller.

39. Switching Coins

The value of the 12 coins is 86 cents.

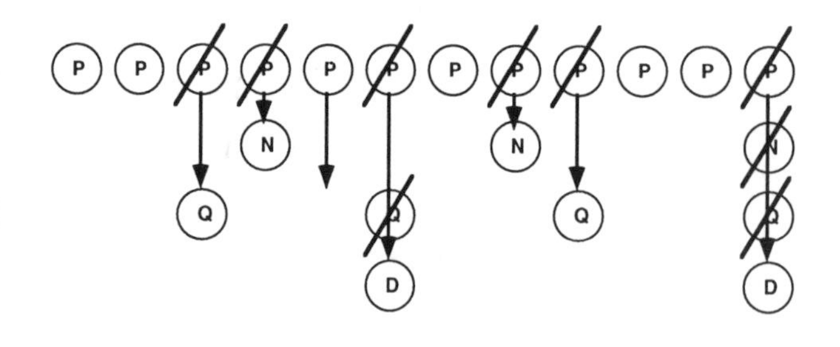

GEOMETRY

40. Counting Squares and Rectangles

There are 5 squares. There are 9 rectangles.

41. More Rectangles

There are 15 rectangles.

42. Counting Triangles

There are 12 triangles.

43. More Counting Triangles

There are 28 triangles.

44. The Missing Perimeter

Since the area of the square is 36 sq. cm. (2×18 sq. cm.), each side is 6 cm. and the perimeter is 4×6 cm., or 24 cm.

45. The Quadrilaterals Go On Stage!

The various quadrilaterals can go in the doors as follow. Door 1: trapezoid only. Door 2: rhombus and square. Door 3: rectangle, parallelogram, rhombus, and square. Door 4: rectangle and square. Door 5: square only.

Different kinds of quadrilaterals can share some of the same characteristics, such as right angles and parallel sides.

46. The Sum of Two Sides

With sides of 5 and 6, the third side can be 10, 9, 8, 7, 6, 5, 4, 3, or 2. It could not be 11 or more because it would equal (or exceed) the sum of the other two sides. It could not be 1 because two sides (5 and 1) would not equal more than the third side (6).

47. The Land Must Be Divided

The land can be divided like this:

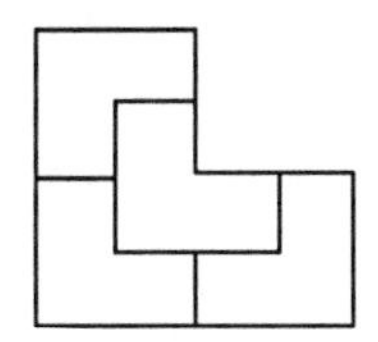

48. Types of Triangles

Equilateral triangles: 1—BED

Scalene triangles: 4—ABD, AFE, FEB

Isosceles triangles: 3—ABC, AEB, BED (the equilateral triangle is also an isosceles triangle).

Remember that equilateral triangles have 3 equal sides, scalene triangles have no equal sides, and isosceles triangles have 2 equal sides.

49. Shadow's Pen

You can make a pen with an area of 625 square yards by making a square fence that is 25 yards by 25 yards.

A circular pen would give the greatest area: Circumference is diameter times pi ($C = \pi d$), so $100 = \pi d$; dividing 100 by pi (about 3.14) gives a diameter of 31.8. The total area (πr^2) is about 793.8 sq. yds.

One reason why many Africans and Native Americans have built circular dwellings is because they could make a larger enclosure with fewer materials.

50. Cutting Circles

Cuts:	1	2	3	4	5	6	7	8
Sections:	2	4	7	11	16	22	29	37

To get from 2 sections to 4 sections, you add 2, then from 4 sections to 7 sections, add 3, from 7 sections to 11 sections, add 4, etc.

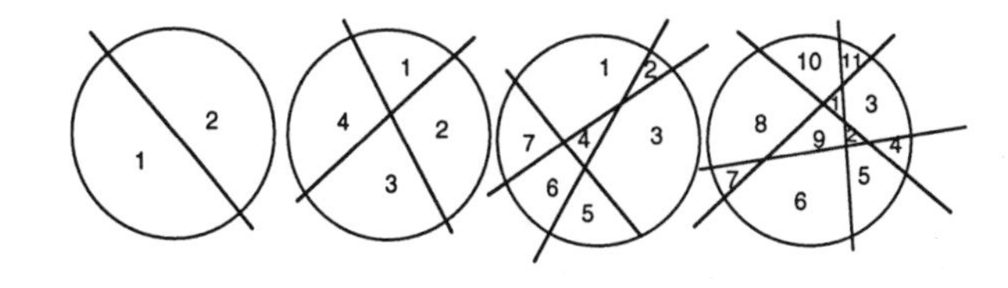

51. The Paint Fell on the Cube

The cubes are painted as follows:

One face painted red: 6. Two faces painted red: 12. Three faces painted red: 8. No faces painted red: 1.

RATIO, PROPORTION AND PERCENT

52. The Bird Shop

Of the 45 birds, there are 18 parrots. You know that 3 canaries plus 2 parrots makes a group of 5. In 45, there are 9 groups of 5. Therefore, there are 3 canaries times 9, or 27 canaries and 2 parrots times 9, or 18 parrots.

53. The Birthday Party

There were 20 girls and 15 boys (5 more girls than boys). The chart below shows how to arrive at the answer:

girls:	4	8	12	16	20
boys:	3	6	9	12	15

54. The Wedding Cake

You need two 24" × 24" cakes. The easiest way to figure this out is to determine how many 8" × 8" cakes fit into a 24" × 24" cake. That's 9 cakes ($8 \times 8 = 64$ and $24 \times 24 = 576$; $576 \div 64 = 9$). Since each 8" × 8" cake serves 8 people, a 24" × 24" cake serves 72 people ($8 \times 9 = 72$). Therefore, it takes two 24" × 24" cakes to serve 144 people.

55. Percent City

Halfway between floors 4 and 5 in Five Town would be the same as the 90th floor (90% of the building height) in Percent City.

56. Is It Free?

No, it's not free. First it's $50, then it's $25 ($50 minus .50 × $50, or $50 minus $25), then $12.50 ($25 minus .50 × $25, or $25 minus $12.50).

57. The Growing Baby

He now weighs 8 lbs. 10 ounces: 8 lbs. – 2 lbs. 4 oz. = 5 lbs. 12 oz. Add 2 lbs. 14 oz. to get 8 lbs. 10 oz.

58. The Sales Tax

Both are right. Multiplying by .08 and adding it to the original amount is the same as multiplying the original amount by 1.08 (since the original amount is 100%, or 1). Instead of taking 8% of 100%, you can just multiply the original amount by 108%, or 1.08.

PROBABILITY

59. Roll One Die

Answers will vary. Students should realize that the outcomes are equally likely.

60. Roll Two Dice

Answers will vary. The outcomes are no longer equally likely, as they were in the previous problem. The sum 7 has more chances of happening.

61. Let's Sit Together

There are 6 ways the three could sit together (N = Nadia, T = Tobi, and E = Elena):

NTE, NET, ENT, ETN, TEN, TNE

62. Amanda's Name

Here is a way Mark could have gotten 9 points: 5 + 1 + 1 + 1 + 1. (Don't forget that Mark always puts the previous letters back, so they can be repicked.)

The A is most likely to be picked since there are three of them in the bag.

63. The Socks Problem

He needs to grab two more socks to be sure of obtaining a match.

64. The Cayuga Indians

The odds of getting 6 black or 6 white or 5 black/1 white or 5 white/1 black are small. It's important to realize that the combinations including 5 out of 6 or 6 out of 6 of one color are harder to get than other combinations. (There are more ways of getting 3 out of 6 or 4 out of 6.)

65. The Family Reunion

There were 21 members altogether. First, write the number 4 (for the four people who are common to both families) in the center. Then subtract 10 – 4 = 6 for the area on the left (Smith side) and 15 – 4 = 11 for the area on the right (Perez side). Check, by adding, to see that your answers are consistent with the information in the problem.

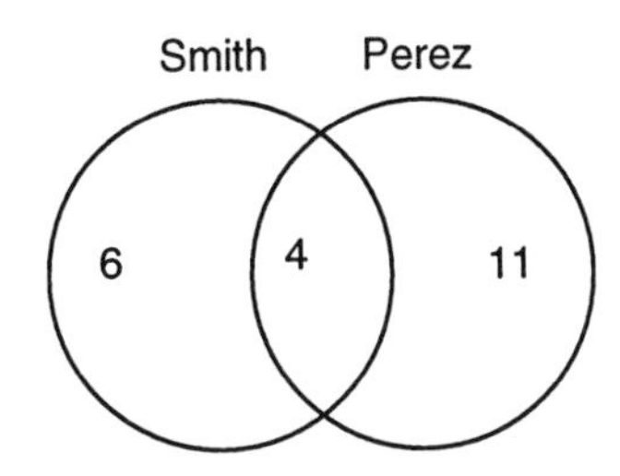

66. Pine Elementary School

There are 81 students participating in clubs. (Use a strategy similar to the one in problem 65.)

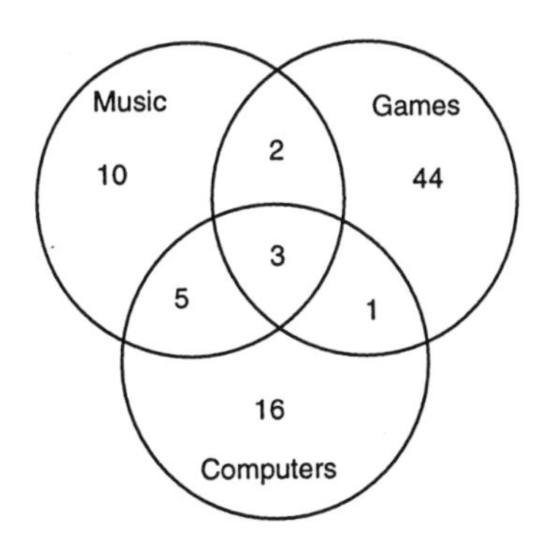

67. The Marble Collection

No, Lucia has fewer than 43 marbles—30, to be exact.

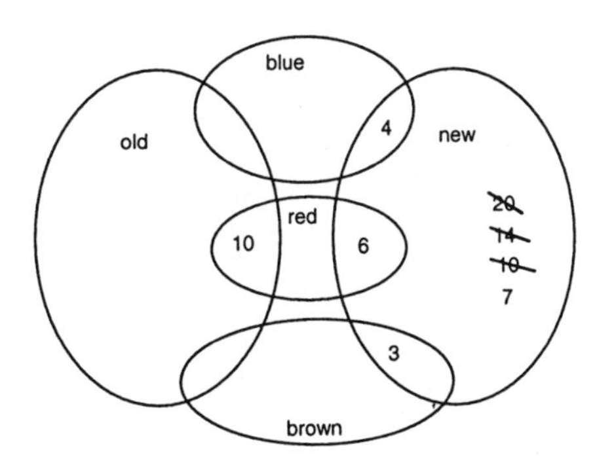

68. Make Your Own Sets Problem

Answers will vary.

PRE-ALGEBRA

69. A Calculator Mistake

The answer is 3. Divide by 10 to undo the mistake and get back to the original number; then divide by 10 again. ($300 \div 10 = 30$; $30 \div 10 = 3$)

70. Another Calculator Mistake

The answer is 36,000. If he divided by 60 and got 10, the original number must have been 600 ($10 \times 60 = 600$). $600 \times 60 = 36{,}000$

71. What values?

The values 0 and 2 work for $x + x = x \bullet x$ ($2 + 2$ is the same as 2×2; $0 + 0$ is the same as 0×0).

All values of x work for $x + x + x = 3 \bullet x$.

72. Three Times My Age?

Mom was right because the ratio of ages changes over time. The following chart shows how Mom's age changes in relation to Jimmy's age:

Mom's age	35	40	45	50	60
Jimmy's age	5	10	15	20	30
Times Mom's age is greater	7	4	3		2

73. Consecutive Whole Numbers

The whole numbers 124, 125, and 126 add up to 375. You can take 375 and divide it by 3 so that 125 + 125 + 125 = 375. However, since the numbers are consecutive, you must take one away from the first 125 and add it to the last 125! Therefore, 124 + 125 + 126 = 375.

You could also figure this out by using algebra: We can call the first number x. Since the numbers are consecutive, the other two are $x + 1$ and $x + 2$. Since the 3 numbers add up to 375, we can say $x + x + 1 + x + 2 = 375$.

That's the same as saying $3x + 3 = 375$. Take 3 from both sides to get $3x = 372$. Then divide both sides by 3 to get $x = 124$. Now we know that $x + 1 = 125$ and $x + 2 = 126$.

74. Balance It!

There are 4 blocks in the bag. If we remove 2 blocks from each side, we see that what is in the bag must equal 4 blocks.

75. Balance It Again!

There are 3 blocks in each bag. Remove two blocks from both sides, remembering to keep the balance. This leaves one bag plus 3 blocks on the left side and only 2 bags on the right side. Remove one bag from each side, leaving 3 blocks on the left side and one bag on the right side. The bag must have 3 blocks!

You could also use algebra: If we call the number of blocks in one bag x, then the x combined with 5 blocks on one side must equal 2 x's plus two blocks on the other side. We can say $x + 5 = 2x + 2$. To keep the equation true, we can subtract 2 from both sides, getting $x + 3 = 2x$. Then we can take one x away from both sides to get $3 = x$.

76. Make Your Own Story

Answers will vary.

77. The Function Machines Are Running Wild!

Each machine takes in a number on the left and spits out a new number on the right. The job each machine performs on its number can be described in the following chart (*n* stands for the number that is fed to the machine).

Machine 1: $2n + 1$

Machine 2: $3n$

Machine 3: $2n - 1$

Machine 4: $3n - 1$

Machine 5: $n/2 - 1$

Machine 6: $n + 3$

MORE CHARTS AND POPULAR PROBLEMS

78. Cutting Up

You have to make 5 cuts.

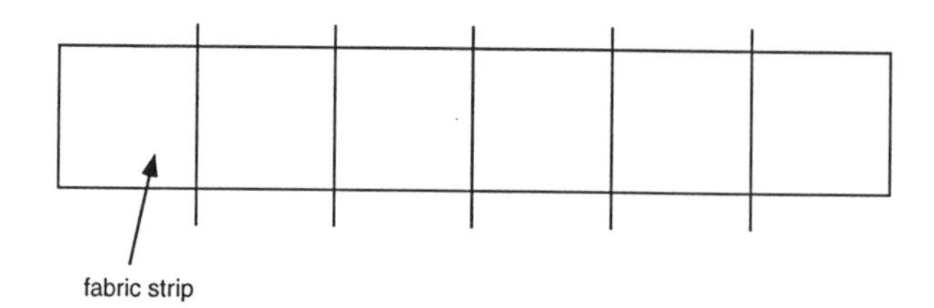

79. The Clock Strikes Again

There are five one-second intervals between strikes 1 and 6, so it takes five seconds between strikes. Likewise, between the first and twelfth strikes, there would be 11 one-second intervals, or 11 seconds. Note: a *strike* is completed instantly (even though the sound of the chime may continue). Therefore, the interval *between* strikes starts at the beginning of the sound of the first chime and ends at the beginning of the last chime.

80. Let's Sit In A Circle

Person 10 sits across from person 3. The best hint is to look at a clock then adjust it for 14 intervals:

81. The Highway

Each lane is 12 3/4 feet or 12 feet 9 inches wide. Subtract 2×12 feet (for the shoulders) from the 78 foot highway to get $78 - 24 = 54$. Subtract 3 feet from the remaining 54 feet to get $54 - 3 = 51$. Divide the remaining 51 feet by 4 lanes to get the width of one lane: $51 \div 4 = 12\ 3/4$, or 12 feet 9 inches.

82. A Sign For Your Name

Answers will vary. Multiply the number

of letters in the name by 1 inch; multiply the spaces between letters by 1/2 inch. Add the results and subtract from 35 inches. To determine the space left on either side of the name, divide by 2. (A name with 3 letters would start 15 1/2 inches from the left side of the sign.)

83. Visiting Your Friend

There are 6 ways to get to your friend's house (don't forget to count every path—going *with* the arrows only).

84. Can You Walk Through Paper?

If the paper is cut on the given lines, students should be able to spread the borders enough to walk through!

85. A Coloring Problem

The given map should take at least 4 colors.

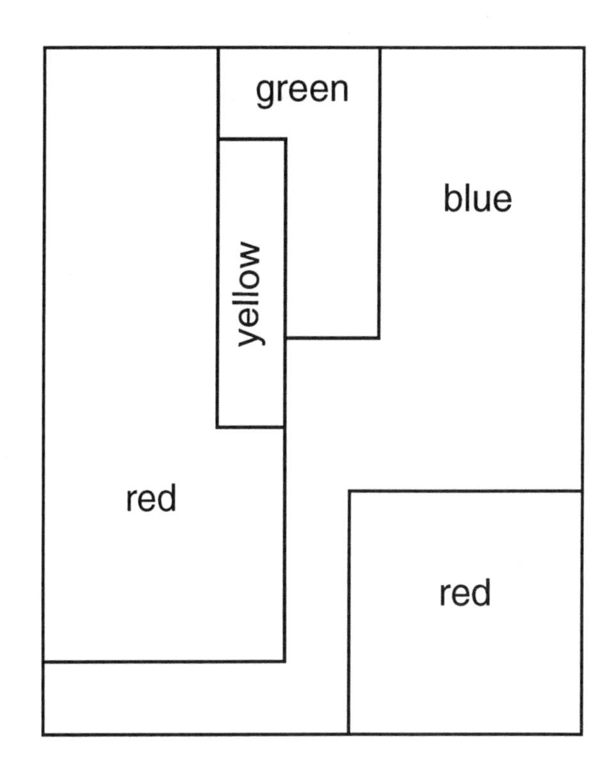

Appendix A

PROBLEM–SOLVING PORTFOLIO

Note to the teacher:

The Problem-Solving Portfolio can be used as an aid to self-analysis, encouraging students to take time to reflect on their problem-solving growth. It can also be used as an evaluation tool, but be careful not to make the portfolio itself the goal of problem solving.

Read the directions to the students, and provide time for students to answer the questions individually. Students should attach the appropriate problem to each questionnaire and make an attractive cover. Allow students to include problems they got wrong, if they wish, as long as they correct the original errors. The emphasis should be on effort.

A rubric for evaluating the Problem-Solving Portfolio has been provided for your convenience.

PROBLEM-SOLVING PORTFOLIO

DIRECTIONS:

I. Choose three problems you have worked on according to the following categories.

1. Most fun
2. Most challenging
3. Problem that best demonstrates your use of problem-solving strategies

NOTE: All problems should have the correct work and answer.

II. Fill out the questionnaires neatly and in complete sentences.

III. Attach each questionnaire to the appropriate problem.

IV. Make an attractive cover. Label it with "PROBLEM-SOLVING PORTFOLIO," your name, and your class period.

MOST FUN PROBLEM

Name of the problem ______________________________

What made this problem so much fun? ______________________________

____

What problem-solving strategies did you use? Explain. ______________________________

____

Was there anything you tried that did not work? If yes, explain below.

I tried this, but it didn't work:	Why it didn't work:

Was there anything you tried that did work?

I tried this and it worked:	Why it worked:

Was this a good problem to do in a group? Why?

____

____

MOST CHALLENGING PROBLEM

Name of the problem ______________________________

What made this problem the most challenging (hardest)? ______________________________

____

What problem-solving strategies did you use? ______________________________

____

What did you try that didn't work?

I tried this, but it didn't work:	Why it didn't work:

What did you try that did work? Explain below:

I tried this and it worked:	Why it worked:

Did it help you to work in a group? Explain why:

____

____

PROBLEM THAT BEST DEMONSTRATES THE USE OF A PROBLEM-SOLVING STRATEGY

Name of the problem ______________________________

What specific strategy or strategies did you use that really helped you? List them below and explain why (it's okay if you used only one):

STRATEGY	HOW IT HELPED

Do you consider yourself a good problem solver? Why or why not? ______________________________

____

What do you think you are best at?

	check here
Being a good team member.	
Reading the problem many times.	
Not quitting.	
Making a drawing or chart.	
Being careful with my work.	
Other:	

How do you think you have improved at problem solving? ______________________________

____

RUBRIC FOR PROBLEM-SOLVING EVALUATION

NAME ______________________________ DATE __________

__________ Choice of problems according to categories (1–3)

__________ Questionnaire—effort, complete sentences, neatness (1–5)

__________ Effort shown on problems (1–5)

__________ Cover (1–2)

__________ Total points (15 total)

__________ **Portfolio grade**

Comments ______________________________

____

____

____

Appendix B

MATH MYTHS

Read each of the following and answer honestly. Check each statement that you agree with. Be ready to support your answers.

_____ 1. Mathematics ability is inherited.

_____ 2. You don't need to study for math.

_____ 3. Reading, writing, and spelling are not important in mathematics.

_____ 4. Boys are better than girls in math or science.

_____ 5. Problem solving is the hardest part of mathematics.

_____ 6. If you don't know how to solve a problem after you read it, you probably don't know enough to solve it.

_____ 7. If you don't understand something in math, it usually means you're not good in math.

_____ 8. If you ask questions in math class, people will think that you are either a "nerd" or just "slow."

_____ 9. Math is easy only when the teacher is reviewing old stuff.

_____ 10. Taking notes in math class is not as important as taking notes in other subjects.

_____ 11. Only those gifted in math or science are able to become mathematicians or scientists.

_____ 12. Mathematicians and scientists don't enjoy sports or other subjects as much as the rest of the population does.

Appendix C

MATRIX OF PROBLEM-SOLVING CONCEPTS

	PROBLEM NUMBER																
CONCEPTS	1	2	3	4	5	6	7	8	9	10	11	12	13	14	15	16	17
Whole Numbers & No. Theory																	
Basic Operations		✓			✓			✓	✓	✓	✓	✓	✓	✓	✓	✓	✓
Order of Operations								✓				✓	✓		✓		
Exponents																	
Factors & Multiples																	
Divisibility																	
Perfect Squares																	
Measurement Conversions													✓				
Fractions																	
General Concepts																	
Addition/Subtraction																	
Multiplication/Division																	
Decimals																	
Place Value																	
Addition/Subtraction																	
Multiplication/Division																	
Ratio																	
Percent Concepts																	
Percent of a Number																	
Geometry																	
Vocabulary							✓										
Quadrilateral Properties																	
Triangle Properties																	
Perimeter																	
Area																	
Volume Concepts																	
Pre-Algebra																	
Consecutive Integers																	
Equation Investigation																	
Working with Variables																	
Function Machines																	
Probability																	
Sets																	
Topology																	
Logic	✓	✓	✓	✓	✓	✓	✓										

MATRIX OF PROBLEM-SOLVING CONCEPTS

	PROBLEM NUMBER																
CONCEPTS	18	19	20	21	22	23	24	25	26	27	28	29	30	31	32	33	34
Whole Numbers & No. Theory																	
Basic Operations			✓	✓		✓	✓	✓							✓		
Order of Operations																	
Exponents								✓									
Factors & Multiples	✓	✓	✓														
Divisibility							✓										
Perfect Squares					✓	✓											
Measurement Conversions		✓											✓		✓		
Fractions																	
General Concepts											✓				✓		
Addition/Subtraction													✓	✓			
Multiplication/Division									✓	✓		✓		✓			
Decimals																	
Place Value																✓	✓
Addition/Subtraction																	
Multiplication/Division																	✓
Ratio																	
Percent Concepts																	
Percent of a Number																	
Geometry																	
Vocabulary																	
Quadrilateral Properties																	
Triangle Properties																	
Perimeter																	
Area					✓												
Volume Concepts																	
Pre-Algebra																	
Consecutive Integers																	
Equation Investigation																	
Working with Variables																	
Function Machines																	
Probability																	
Sets																	
Topology																	
Logic																	

MATRIX OF PROBLEM-SOLVING CONCEPTS

	PROBLEM NUMBER																
CONCEPTS	35	36	37	38	39	40	41	42	43	44	45	46	47	48	49	50	51
Whole Numbers & No. Theory																	
Basic Operations			✓														
Order of Operations																	
Exponents																	
Factors & Multiples																	
Divisibility																	
Perfect Squares																	
Measurement Conversions				✓													
Fractions																	
General Concepts	✓																
Addition/Subtraction																	
Multiplication/Division	✓																
Decimals																	
Place Value																	
Addition/Subtraction		✓			✓												
Multiplication/Division		✓	✓		✓												
Ratio																	
Percent Concepts																	
Percent of a Number																	
Geometry																	
Vocabulary						✓	✓	✓	✓				✓	✓			
Quadrilateral Properties											✓						
Triangle Properties												✓					
Perimeter										✓							
Area										✓					✓	✓	
Volume Concepts																	✓
Pre-Algebra																	
Consecutive Integers																	
Equation Investigation																	
Working with Variables																	
Function Machines																	
Probability																	
Sets																	
Topology																	
Logic																	

MATRIX OF PROBLEM-SOLVING CONCEPTS

	PROBLEM NUMBER																
CONCEPTS	52	53	54	55	56	57	58	59	60	61	62	63	64	65	66	67	68
Whole Numbers & No. Theory																	
Basic Operations																	
Order of Operations																	
Exponents																	
Factors & Multiples																	
Divisibility																	
Perfect Squares																	
Measurement Conversions																	
Fractions																	
General Concepts																	
Addition/Subtraction																	
Multiplication/Division																	
Decimals																	
Place Value																	
Addition/Subtraction																	
Multiplication/Division																	
Ratio	✓	✓	✓														
Percent Concepts				✓	✓	✓	✓										
Percent of a Number					✓	✓	✓										
Geometry																	
Vocabulary																	
Quadrilateral Properties																	
Triangle Properties																	
Perimeter																	
Area																	
Volume Concepts																	
Pre-Algebra																	
Consecutive Integers																	
Equation Investigation																	
Working with Variables																	
Function Machines																	
Probability								✓	✓	✓	✓	✓	✓				
Sets														✓	✓	✓	✓
Topology																	
Logic																	

MATRIX OF PROBLEM-SOLVING CONCEPTS

	PROBLEM NUMBER																
CONCEPTS	69	70	71	72	73	74	75	76	77	78	79	80	81	82	83	84	85
Whole Numbers & No. Theory																	
Basic Operations	✓	✓											✓				
Order of Operations																	
Exponents																	
Factors & Multiples																	
Divisibility																	
Perfect Squares																	
Measurement Conversions																	
Fractions																	
General Concepts														✓			
Addition/Subtraction														✓			
Multiplication/Division														✓			
Decimals																	
Place Value																	
Addition/Subtraction																	
Multiplication/Division																	
Ratio				✓													
Percent Concepts																	
Percent of a Number																	
Geometry																	
Vocabulary																	
Quadrilateral Properties																	
Triangle Properties																	
Perimeter																	
Area																	
Volume Concepts																	
Pre-Algebra																	
Consecutive Integers					✓												
Equation Investigation	✓	✓				✓	✓	✓									
Working with Variables			✓														
Function Machines									✓								
Probability																	
Sets																	
Topology																✓	✓
Logic										✓	✓	✓			✓		

Appendix D

MATRIX OF PROBLEM-SOLVING STRATEGIES

STRATEGIES	PROBLEM NUMBER																				
	1	2	3	4	5	6	7	8	9	10	11	12	13	14	15	16	17	18	19	20	21
Make a diagram, chart, or table	✓	✓		✓	✓	✓	✓	✓	✓	✓	✓	✓	✓	✓	✓	✓	✓	✓	✓	✓	✓
Try easier numbers																					
Find a pattern or formula											✓					✓	✓	✓	✓	✓	✓
Act it out!	✓	✓	✓	✓	✓	✓								✓			✓				
Use manipulatives	✓	✓		✓	✓	✓	✓		✓		✓			✓				✓			✓
Work backwards																					
Guess and check				✓		✓		✓		✓		✓									
Explore other answers	✓	✓			✓			✓		✓											

STRATEGIES	PROBLEM NUMBER																				
	22	23	24	25	26	27	28	29	30	31	32	33	34	35	36	37	38	39	40	41	42
Make a diagram, chart, or table	✓	✓	✓	✓	✓	✓	✓		✓	✓	✓			✓	✓		✓	✓	✓	✓	✓
Try easier numbers					✓		✓						✓			✓					
Find a pattern or formula	✓	✓	✓	✓																✓	
Act it out!																					
Use manipulatives	✓			✓	✓	✓	✓				✓	✓						✓			
Work backwards								✓								✓					
Guess and check			✓											✓	✓						
Explore other answers							✓														

MATRIX OF PROBLEM-SOLVING STRATEGIES

STRATEGIES	PROBLEM NUMBER																				
	43	44	45	46	47	48	49	50	51	52	53	54	55	56	57	58	59	60	61	62	63
Make a diagram, chart, or table	✓	✓	✓	✓	✓	✓	✓	✓	✓	✓	✓	✓	✓	✓	✓	✓	✓	✓	✓		✓
Try easier numbers																					
Find a pattern or formula							✓	✓		✓	✓								✓		
Act it out!			✓														✓	✓	✓	✓	✓
Use manipulatives		✓			✓				✓			✓	✓	✓			✓	✓	✓	✓	✓
Work backwards		✓																			
Guess and check				✓			✓			✓	✓										
Explore other answers							✓										✓	✓		✓	

STRATEGIES	PROBLEM NUMBER																					
	64	65	66	67	68	69	70	71	72	73	74	75	76	77	78	79	80	81	82	83	84	85
Make a diagram, chart, or table	✓	✓	✓	✓	✓			✓	✓	✓	✓	✓		✓	✓	✓	✓	✓	✓	✓		✓
Try easier numbers						✓				✓												
Find a pattern or formula									✓	✓				✓			✓					
Act it out!	✓												✓		✓		✓			✓		
Use manipulatives	✓	✓	✓	✓	✓						✓	✓	✓		✓		✓		✓		✓	
Work backwards						✓	✓				✓	✓	✓									
Guess and check						✓	✓	✓	✓	✓	✓	✓	✓									
Explore other answers								✓						✓								✓